Computer Aided Optimum Design in Engineering XI

WIT*PRESS*

WIT Press publishes leading books in Science and Technology.
Visit our website for new and current list of titles.
www.witpress.com

WIT*eLibrary*

Home of the Transactions of the Wessex Institute.
Papers presented at OPTI 2009 are archived in the WIT eLibrary in volume 106 of WIT Transactions on The Built Environment (ISSN 1743-3509).
The WIT eLibrary provides the international scientific community with immediate and permanent access to individual papers presented at WIT conferences.
Visit the WIT eLibrary at www.witpress.com.

ELEVENTH INTERNATIONAL CONFERENCE ON OPTIMUM DESIGN OF STRUCTURES AND MATERIALS IN ENGINEERING

OPTI XI

CONFERENCE CHAIRMEN

C.A. Brebbia
Wessex Institute of Technology, UK

S. Hernández
University of La Coruña, Spain

INTERNATIONAL SCIENTIFIC ADVISORY COMMITTEE

Organised by
Wessex Institute of Technology, UK
&
University of La Coruña, Spain

Sponsored by
WIT Transactions on The Built Environment

WIT Transactions

Transactions Editor

Editorial Board

S Lomov Katholieke Universiteit Leuven, Belgium

J W S Longhurst University of the West of England, UK

G Loo The University of Auckland, New Zealand

J Lourenco Universidade do Minho, Portugal

J E Luco University of California at San Diego, USA

H Lui State Seismological Bureau Harbin, China

C J Lumsden University of Toronto, Canada

L Lundqvist Division of Transport and Location Analysis, Sweden

T Lyons Murdoch University, Australia

Y-W Mai University of Sydney, Australia

M Majowiecki University of Bologna, Italy

D Malerba Università degli Studi di Bari, Italy

G Manara University of Pisa, Italy

B N Mandal Indian Statistical Institute, India

Ü Mander University of Tartu, Estonia

H A Mang Technische Universitat Wien, Austria,

G D, Manolis, Aristotle University of Thessaloniki, Greece

W J Mansur COPPE/UFRJ, Brazil

N Marchettini University of Siena, Italy

J D M Marsh Griffith University, Australia

J F Martin-Duque Universidad Complutense, Spain

T Matsui Nagoya University, Japan

G Mattrisch DaimlerChrysler AG, Germany

F M Mazzolani University of Naples "Federico II", Italy

K McManis University of New Orleans, USA

A C Mendes Universidade de Beira Interior, Portugal,

R A Meric Research Institute for Basic Sciences, Turkey

J Mikielewicz Polish Academy of Sciences, Poland

N Milic-Frayling Microsoft Research Ltd, UK

R A W Mines University of Liverpool, UK

C A Mitchell University of Sydney, Australia

K Miura Kajima Corporation, Japan

A Miyamoto Yamaguchi University, Japan

T Miyoshi Kobe University, Japan

G Molinari University of Genoa, Italy

T B Moodie University of Alberta, Canada

D B Murray Trinity College Dublin, Ireland

G Nakhaeizadeh DaimlerChrysler AG, Germany

M B Neace Mercer University, USA

D Necsulescu University of Ottawa, Canada

F Neumann University of Vienna, Austria

S-I Nishida Saga University, Japan

H Nisitani Kyushu Sangyo University, Japan

B Notaros University of Massachusetts, USA

P O'Donoghue University College Dublin, Ireland

R O O'Neill Oak Ridge National Laboratory, USA

M Ohkusu Kyushu University, Japan

G Oliveto Universitá di Catania, Italy

R Olsen Camp Dresser & McKee Inc., USA

E Oñate Universitat Politecnica de Catalunya, Spain

K Onishi Ibaraki University, Japan

P H Oosthuizen Queens University, Canada

E L Ortiz Imperial College London, UK

E Outa Waseda University, Japan

A S Papageorgiou Rensselaer Polytechnic Institute, USA

J Park Seoul National University, Korea

G Passerini Universita delle Marche, Italy

B C Patten, University of Georgia, USA

G Pelosi University of Florence, Italy

G G Penelis, Aristotle University of Thessaloniki, Greece

W Perrie Bedford Institute of Oceanography, Canada

R Pietrabissa Politecnico di Milano, Italy

H Pina Instituto Superior Tecnico, Portugal

M F Platzer Naval Postgraduate School, USA

D Poljak University of Split, Croatia

V Popov Wessex Institute of Technology, UK

H Power University of Nottingham, UK

D Prandle Proudman Oceanographic Laboratory, UK

Computer Aided Optimum Design in Engineering XI

EDITORS

S. Hernández
University of La Coruña, Spain

C.A. Brebbia
Wessex Institute of Technology, UK

WIT*PRESS* Southampton, Boston

Editors:

S. Hernández
University of La Coruña, Spain

C.A. Brebbia
Wessex Institute of Technology, UK

Published by

WIT Press
Ashurst Lodge, Ashurst, Southampton, SO40 7AA, UK
Tel: 44 (0) 238 029 3223; Fax: 44 (0) 238 029 2853
E-Mail: witpress@witpress.com
http://www.witpress.com

For USA, Canada and Mexico

Computational Mechanics Inc
25 Bridge Street, Billerica, MA 01821, USA
Tel: 978 667 5841; Fax: 978 667 7582
E-Mail: info@compmech.com
http://www.witpress.com

British Library Cataloguing-in-Publication Data

A Catalogue record for this book is available
from the British Library

ISBN: 978-1-84564-185-6
ISSN: 1746-4498 (print)
ISSN: 1743-3509 (on-line)

The texts of the papers in this volume were set individually by the authors or under their supervision. Only minor corrections to the text may have been carried out by the publisher.

Printed in Great Britain by the MPG Books Group, Bodmin and King's Lynn

Preface

OPTI 2009, the eleventh edition in the series of international conferences devoted to Computer Aided Optimum Design of Structures was held in the Algarve (Portugal) organized by the Wessex Institute of Technology, U.K. and the University of La Coruña, Spain. The main objective of the event was to provide a friendly and useful forum for the interchange of ideas and initiatives amongst researchers, designers, and scholars of the optimization community to share advances in design optimization. The OPTI Conferences started in 1989, when it took place in Southampton, U.K. After that every two years a new meeting has been held alternatively at each side of the Atlantic Ocean, in locations such as Boston, USA (1991); Zaragoza, Spain (1993); Miami, USA (1995); Rome, Italy (1997); Orlando, USA (1999); Bologna, Italy (2001); Detroit, USA (2003); Skiathos, Greece (2005) and Myrtle Beach, USA (2007).

Nowadays, it is already widely accepted that optimization techniques have much to offer to designers involved in the design of new industrial products. The formulation of optimum design has evolved from the time it was an academic topic unable to satisfy the requirements of real life prototypes. Since then the development of new algorithms, the improvement of others, the appearance of powerful commercial computer codes with easy to use graphical interfaces and the revolution in speed of digital computers created a fertile field for the incorporation of optimization in the design process in several engineering disciplines. Although not all industrial sectors are using optimisation techniques daily they are steadily being applied in more design cases.

This book contains most papers presented in OPTI 2009. Following the spirit of previous editions some of them deal with algorithms while others describe innovative design optimization formulations applied in several engineering fields including practical applications in industrial problems.

The content of the book has been distributed into six topics:

- New and enhanced algorithms
- Shape optimization
- Topology optimisation
- Design optimisation in materials, construction and bridge engineering
- Design optimisation in aircraft engineering
- Design optimisation in dam and soil engineering

The wide list of themes shows the maturity of optimization methodologies and their vast number of possible applications. This volume constitutes a valuable addition to the existing literature.

The Editors would like to express their gratitude to all authors for their contributions. Also they are indebted to the members of the International Scientific Advisory Committee of OPTI 2009 who reviewed the manuscripts efficiently and timely, thus ensuring their technical quality.

The Editors
June 2009

Contents

Section 4: Design optimisation in materials, construction and bridge engineering

Section 5: Design optimisation in aircraft engineering

Section 1
New and enhanced algorithms

Application of self-organizing maps to genetic algorithms

S. Kan, Z. Fei & E. Kita
Graduate School of Information Sciences, Nagoya University, Japan

Abstract

This paper describes Self-Organizing Maps for Genetic Algorithm (SOM-GA). In this algorithm, the search performance of a real-coded genetic algorithm (RCGA) is enhanced with self-organizing map (SOM). The SOM is trained with the information of the individuals in the population. Sub-populations are generated from a whole population by the help of the map. The RCGA search is performed in the sub-populations. The Rastrigin function is considered as a test problem. The search performance of SOM-GA is compared with that of the RCGA. The results show that the use of the sub-population search algorithm improves the local search performance of the RCGA and therefore, SOM-GA can find better solutions in shorter CPU time than RCGA.
Keywords: real-coded genetic algorithms, self-organizing maps, Rastrigin function.

1 Introduction

Evolutionary Computations can be classified into Genetic Algorithms (GA)[1, 2], Genetic Programming (GP) [3, 4], Evolutionary Strategy (ES) [5] and so on. The GA, which is one of the most popular evolutionary computations, has been presented by Holland in 1970s. Since then, theoretical study of GA operators and the engineering application of GA have been studied widely by many researchers. Recently, their attention focuses on the multi-objective optimization and real-coded optimization problems from the viewpoint of engineering applications. Now, we will focus on the real-coded optimization problem.

An original genetic algorithm is designed for optimization problems with binary-coded design variables. Therefore, in the previous studies, real-coded design variables have to be transformed to binary-coded ones. However, some researchers

WIT Transactions on The Built Environment, Vol 106, © 2009 WIT Press
www.witpress.com, ISSN 1743-3509 (on-line)
doi:10.2495/OP090011

have presented the real-coded GA (RCGA) in which the real-coded design variables are not transformed into the binary-coded ones [6–8]. In this study, we will present the RCGA with self-organizing maps (SOM), which is named as "Self-Organizing Maps for Genetic Algorithms (SOM-GA)".

The SOM, which has been presented by Kohonen [9] and Van Hulle [10], is a single layer feed-forward network where the output syntaxes are arranged in grid. Simply, the SOM-GA is the combinational algorithm of the Real-Coded Genetic Algorithm (RCGA) and the SOM clustering. The algorithm starts from the definition of the population of individuals. A self-organizing map is trained with the objective function and the design variables of the individuals in the population. The best match unit for a individual is chosen and then, a sub-population is defined by the individuals included in the circle centering on the best match unit. RCGA is performed iteratively in sub-populations and then, the obtained best individuals are added to the next whole population.

The combinational algorithm of the evolutionary algorithm and the SOM has already been presented by Buche *et al.* [11], which is named as Self-Organizing Maps for Multi-Objective Evolutionary Algorithms (SOM-MOEA). The SOM-MOEA is designed for the multi-objective optimization problems and therefore, their aim is different from the SOM-GA.

In the numerical example, Rastrign function is taken as a test function. The SOM-GA and RCGA are compared in their search performance by finding an optimal solution of the Rastrign function.

2 Self-organizing map

Before explanation of Self-Organizing Maps for Genetic Algorithms (SOM-GA), we will introduce Self-Organizing Maps [9, 10].

The self-organizing map is a single layer feed-forward network where the output syntaxes are arranged in grid (Fig. 1). Each input is connected to all output neurons. A weight vector with the same dimensionality as the input vectors is attached to every neuron. The number of input dimensions is usually a lot higher than the output grid dimension. SOMs are mainly used for dimensionality reduction rather than expansion.

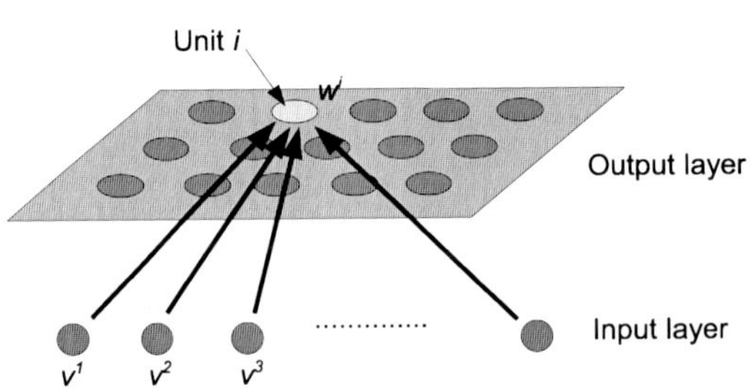

Figure 1: Self-organizing map.

The weight vector w^i at the unit i is given as

$$\boldsymbol{w}^i = \left\{w_1^i, w_2^i, \ldots, w_n^i\right\}^T, \tag{1}$$

and the input vector v^j is as

$$\boldsymbol{v}^j = \left\{v_1^j, v_2^j, \ldots, v_n^j\right\}^T. \tag{2}$$

Taking the Euclid distance $\left\|\boldsymbol{v}^j - \boldsymbol{w}^i\right\|$ as a norm, the best match unit is selected so that the norm is minimized. The best match unit $c\left(\boldsymbol{v}^j\right)$ can be defined as

$$c\left(\boldsymbol{v}^j\right) = \arg\min\{\left\|\boldsymbol{v}^j - \boldsymbol{w}^i\right\|\}. \tag{3}$$

Once determining $c(v^j)$, the weight vector is updated by

$$\boldsymbol{w}^i(t+1) = \boldsymbol{w}^i(t) + h_{ci}(t)\left\{\boldsymbol{v}^j(t) - \boldsymbol{w}^i(t)\right\}, \tag{4}$$

where t denotes the discretized time; $t = 0, 1, 2, \ldots$. The neighborhood function $h_{ci}(t)$ is defined as

$$h_{ci}(t) = \alpha_s \exp\left(-\frac{\left\|r_c - r_i\right\|^2}{2\sigma^2(t)}\right), \tag{5}$$

where r_i and r_c denote the position vectors of the unit i and the best match unit, respectively. The coefficient α_s is sometimes defined as a monotonically decreasing one. In this study, however, it is defined as the constant within the range of $0 < \alpha_s < 1$. The function $\sigma(t)$ is defined as

$$\sigma(t) = \sigma(t-1) - \frac{R}{TS}, \tag{6}$$

where $\sigma(0) = R$ and $\sigma(TS) = 0$. The parameters TS and R denote the number of training step and the initial radius, respectively.

The SOM algorithm is as follows.

1. Initialize randomly the weight vectors $\mathbf{w}^i$.
2. Select the unit $c(\mathbf{v}^j)$ so as to satisfy equation (3).
3. Update the weight vector $\mathbf{w}^i$ according to equation (4).
4. Update the neighborhood function according to equation (5).
5. Repeat from step 2 to step 4.

3 SOM-GA algorithm

For explanation of the SOM-GA algorithm, we will take the following notations.

$\boldsymbol{P}(t)$	Population at generation t.
$p_i(t)$	An individual in a population $\boldsymbol{P}(t)$.
M	Total number of individuals in a population $\boldsymbol{P}(t)$.
$\boldsymbol{SP}_i(t)$	Sub-populations generated from a population $\boldsymbol{P}(t)$.
Map	Self-Organizing Maps.
Rn	Radius for sub-population $\mathbf{SP_i(t)}$ in SOM-GA.

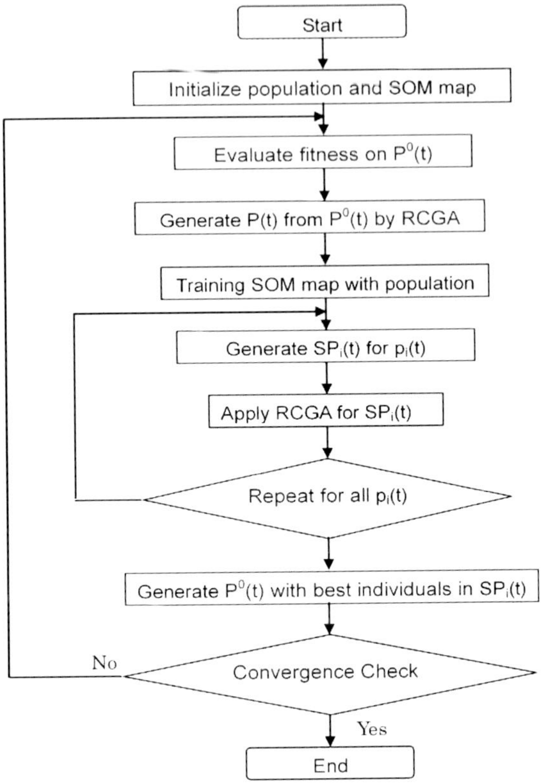

Figure 2: Flow chart of SOM-GA algorithm.

The SOM-GA algorithm (2) is as follows.

1. $t = 0$.
2. Initialize self-organizing map Map.
3. Generate randomly M individuals to construct an initial population $\boldsymbol{P}^0(t)$.
4. Evaluate fitness of individuals.
5. Generate the population $\boldsymbol{P}(t)$ as follows.
    (a) Select two individuals from the population $\boldsymbol{P}^0(t)$ by roulette selection.
    (b) Apply BLX-α crossover [6, 12] to generate new individuals.
    (c) Repeat until generating M individuals.
6. Train Map with the values of the objective function and the design variables of the individuals in the population $\boldsymbol{P}(t)$ according to the Section 3.
7. Generate new individuals as follows.
    (a) Define sub-populations on Map by all individuals within a radius of Rn from the best match unit for the individual $p_i(t)$.
    (b) Perform RCGA operation in the sub-population $\boldsymbol{SP}_i(t)$ to obtaining best individual $p_i(t+1)$.
    (c) Repeat for every individual $p_i(t)(i = 1, 2, \ldots, M)$.

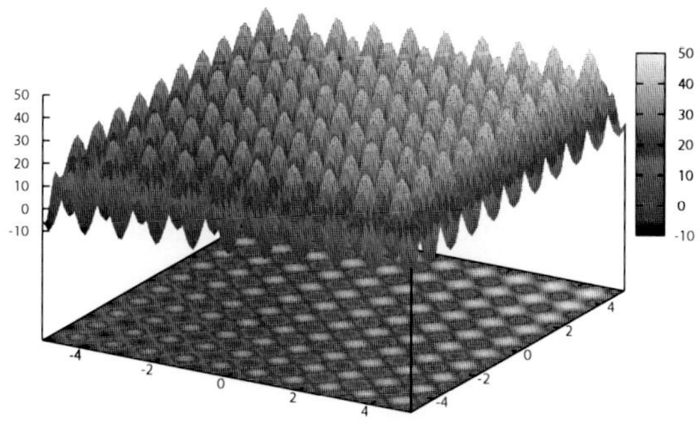

Figure 3: Rastrigin function in two-dimension.

8. Terminate process if convergence criterion is satisfied.
9. Construct new population $\boldsymbol{P}^0(t+1)$ by individuals $p_i(t+1)$ $(i=1,2,\ldots,M)$.
10. Update time step; $t \leftarrow t+1$.
11. Go to (4).

In the numerical examples, the test functions are taken as the objective functions.

4 Numerical examples

4.1 Problem setting

The SOM-GA is applied for finding an optimum solution of Rastrigin function. The test functions are taken as the objective functions and, the SOM map is trained with the values of the test functions and the design variables.

Rastrigin function is defined by

$$F(x) = 10n + \sum_{i=1}^{n} (x_i^2 - 10\cos(2\pi x_i))$$
$$(-5.12 < x_i < 5.12) \tag{7}$$
$$\min(F(x)) = F(0,0,\ldots,0) = 0.$$

In case of two design variables, the function is shown in Fig. 3. Since Rastrigin function is a multi-modal function, it has one global minimum and many local minima. Design variables are independent each other.

In this paper, the number of the design variables is specified as $N_{dv} = 20$.

Common parameters for the SOM-GA and the RCGA are listed in Table 1. Number of the design variables is 20. Population size is 50. Crossover rate and the mutation rate are 1.0 and 0.002, respectively. Roulette selection and BLX-α crossover operations are adopted. In the real-coded mutation operation, the values of the design variables are changed randomly.

Table 1: Common parameters for both algorithms.

Number of design variables	$N_{dv} = 20$
Population size	$N_p = 50$
Selection	Roulette selection
Crossover	BLX-α, $\alpha = 0.5$
Crossover rate	1.0
Mutation rate	0.002
Number of trials	10

Table 2: Additional parameters for SOM-GA.

Map size and type	Hexagon of 20×20
Initial neighborhood radius	$R = 10$
Training rate	$\alpha_s = 0.8$
Training times	$TS = 1000$
Neighborhood radius	$Rn = 2$
Operation number in sub-population	$N_g^{sub} = 1000$

Table 3: Maximum generation step N_g.

	Rastrigin
$(N_g)_{RCGA}$	1,000,000
$(N_g)_{SOM-GA}$	20

Additional parameters for the SOM-GA are shown in Table 2. The neighborhood radius Rn is the radius of the circular region on the self-organizing map which covers a sub-population. The operation number in the subpopulation N_g^{sub} is the number of the RCGA operations in each sub-population.

Ten trials are performed in every cases from different initial populations. Mean values are shown in the numerical results.

Maximum generation steps N_g are shown in Fig. 3. In the SOM-GA, the RCGA operation is performed in every sub-populations. The estimation time of fitness function is equal to the product of the number of maximum generation step, the population size, and the number of the RCGA operations in every sub-populations. In order to equalize the estimation times of fitness functions in RCGA and

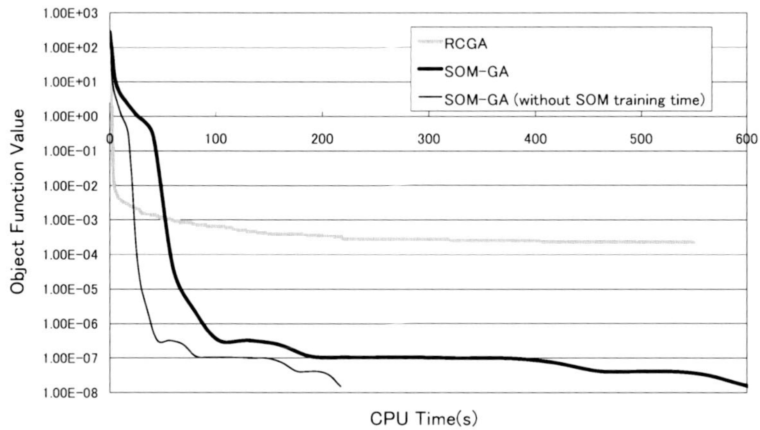

Figure 4: The comparison of RCGA and SOM-GA.

SOM-GA, the generation steps are specified according to

$$(N_g)_{RCGA} \geq (N_g)_{SOM-GA} \times N_p \times N_g^{sub}.$$

where $(N_g)_{RCGA}$ and $(N_g)_{SOM-GA}$ denote the maximum generation step in RCGA and SOM-GA, respectively.

4.2 Numerical results

Convergence histories of the best individuals are shown in Fig. 4. The abscissa and the ordinate denote the CPU time and the objective function of the best individual, respectively. The RCGA and SOM-GA results are labeled with "RCGA" and "SOM-GA", respectively. The convergence speed of RCGA is faster than that of the SOM-GA in the early steps. On the contrary, the convergence speed of SOM-GA is slower than the RCGA in the early steps and then, is accelerated when the objective function value is smaller than 0.001. A final solution is much better than that by RCGA.

The CPU time of the SOM-GA without SOM training time is also shown in the figure. The CPU time of the SOM-GA with SOM training time is twice as long as that without the SOM training time. This means that the computational cost for SOM training is almost equal to the remaining cost.

Next, we would like to discuss the effect of the map size to the search performance. The map size varies from 10×10 to 80×80. The results are compared in Fig. 5. The abscissa and the ordinate denote the dimensionless CPU time with the maximum CPU time of maximum map size and the value of objective function, respectively and the labels denote the map size. We notice that the CPU time increases according to the magnitude of the map size. When the map size is not small, i.e., greater than 30×30, the final solution converges to the same solution. Therefore, we can say in this case that the map size of 30×30 is best from the

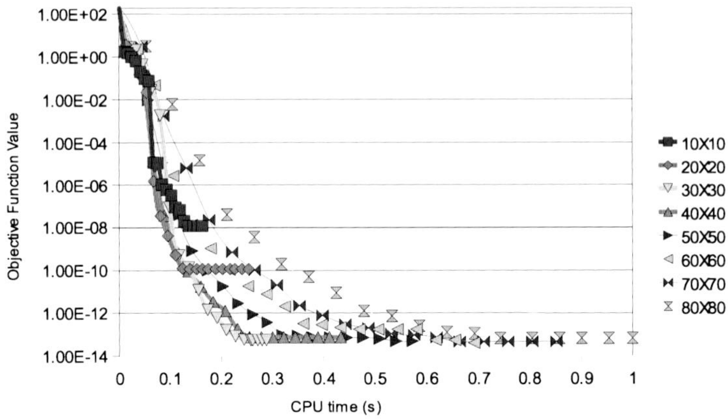

Figure 5: Effect of SOM map-size.

view-points of the search performance and the CPU time. Since the most adequate map size depends on the problem to be solved, we take a small map-size, i.e., the map size of 20×20, in the following examples.

5 Conclusions

This paper described the Self-Organizing Maps for Genetic Algorithms (SOM-GA) for real-valued single objective function problems. In the algorithm, the self-organizing maps are trained with the values of the objective function and the design variables of the individuals in a population and the sub-populations are defined by the help of the SOM clustering. The real-coded genetic algorithm (RCGA) is applied to the individuals in each sub-population. The processes are repeated until satisfying the convergence criterion.

The SOM-GA is applied for finding an optimum solution of Rastrigin function. The results show that the SOM-GA can find better solutions in shorter computational time than the original RCGA without SOM.

References

[1] J. H. Holland. *Adaptation in Natural and Artificial Systems*. The University of Michigan Press, 1 edition, 1975.

[2] D. E. Goldberg. *Genetic Algorithms in Search, Optimization and Machine Learning*. Addison Wesley, 1 edition, 1989.

[3] J. R. Koza, editor. *Genwetic Programming II*. The MIT Press, 1994.

[4] J. R. Koza, F. H. Bennett III, D. Andre, and M. A. Keane, editors. *Genewtic Programming III*. Morgan Kaufmann Pub., 1999.

[5] T. Bäck, editor. *Evolutionary Algorithms in Theory and Practice*. Oxford University Press, 1996.

[6] L. J. Eshelman and J. D. Schaffer. Real-coded genetic algorithms and interval schemata. In L. D. Whitley, editor, *Foundation of Genetic Algorithms 2*, pp. 187–202. Morgan Kaufmann Publications, 1992.

[7] O. Takahashi, H. Kita, and S. Kobayashi. A real-coded genetic algorithm using distance dependent alternation model for complex function optimization. In Darrell Whitley, David Goldberg, Erick Cantu-Paz, Lee Spector, Ian Parmee, and Hans-Georg Beyer, editors, *Proc. of Genetic and Evolutionary Computation Conf. (GECCO2000)*, pp. 219–226, 2000.

[8] H. Kita. A comparison study of self-adaptation in evolution strategies and real-coded genetic algorithms, evolutionary computation. *Evolutionary Computation*, Vol. 9, No. 2, pp. 223–242, 2001.

[9] T. Kohonen. *Self-Organizing Maps*. Springer Verlag, 3 edition, 2001.

[10] M. M. Van Hulle. *Faithful Representations and Topographic Maps*. John Wiley & Sons, 2000.

[11] D. Buche, M. Milano, and P. Koumoutsakos. Self-organizing maps for multi-objective optimization. In A.M. Barry, editor, *GECCO 2002: Proceedings of the Bird of a Feather Workshops, Genetic and Evolutionary Computation Conference, New York, AAAI*, pp. 152–155, 2002.

[12] M. Takahashi and H. Kita. A crossover operator using independent component analysis forreal-coded genetic algorithms. In *Evolutionary Computation*, Vol. 1, pp. 643–649, 2001.

Semi-infinite programming in multi-hazard structural design

F. A. Potra[1] & E. Simiu[2]
[1]*Mathematical and Computational Sciences Division, National Institute of Standards and Technology, Gaithersburg, MD, USA*
[2]*Materials and Construction Research Division, National Institute of Standards and Technology, Gaithersburg, MD, USA*

Abstract

Optimal multi-hazard structural design consists in determining a vector of design variables, subjected to the constraints imposed by all the hazards to which the structure is exposed, such that the cost or weight of the structure is minimized. In particular, the design variables should be determined is such a way that the load-induced stresses and deflections are kept below specified thresholds at all points of the resulting structure. Since there are infinitely many such points, the optimization problem becomes a semi-infinite programming problem. In the present paper we discuss the difficulties involved in the numerical solution of the semi-infinite programming problems arising in multi-hazard structural design. We show that it is possible to construct efficient and robust optimization algorithms, by adaptively choosing a family of finite sets of points on the structure, and by using interior point methods for solving the corresponding optimization problems.

1 Introduction

Optimal structural design consists in determining a vector $d = (d_1, d_2, \ldots, d_n)$ of n design variables such that the structure defined by d satisfies a set of given requirements, and a certain objective function, $f(d)$, related to the cost of the structure is minimized. Multi-hazard design is required where the structure may be subjected to p different hazards. Some of the requirements to be satisfied by the design variables are independent of the different hazards, and can be expressed as inequality constraints of the form $g(d) \leq 0$, where $g : I\!R^n \rightarrow I\!R^m$ is a given vector function. The inequality $g(d) \leq 0$ is interpreted component-wise,

WIT Transactions on The Built Environment, Vol 106, © 2009 WIT Press
www.witpress.com, ISSN 1743-3509 (on-line)
doi:10.2495/OP090021

so that the design variables are subject to m constraints that are independent of the hazards. For example, some of these m constraints are simple bound constraints, imposing lower and upper bounds for the values of the design variables. On the other hand, each hazard i, $i = 1, 2, \ldots, p$, imposes constraints of the form $g^i(d, h) \leq 0, \forall h \in \mathcal{H}$, where $g^i : I\!R^n \times \mathcal{H} \to I\!R^{m_i}$. Here $\mathcal{H}$ is an infinite family of points belonging to the structure. The constraint $g^i(d, h) \leq 0, \forall h \in \mathcal{H}$ specifies that the load-induced stresses and deflections caused by hazard i are kept below specified thresholds at all points of $\mathcal{H}$. With this notation, the optimal multi-hazard structural design can be written as

$$\begin{aligned} &\min f(d) \\ \text{such that } &g(d) \leq 0\,, \\ &g^1(d,h) \leq 0, \ \ \forall h \in \mathcal{H}\,, \\ &\ldots \\ &g^p(d,h) \leq 0, \ \ \forall h \in \mathcal{H}\,. \end{aligned} \tag{1}$$

Since $\mathcal{H}$ is an infinite set, the above optimization problem is a semi-infinite optimization problem [1–3]. Such problems are very difficult to solve, because they contain an infinite number of constraints. A simple minded approach is to replace in (1) the infinite set $\mathcal{H}$ by a finite subset $\widetilde{\mathcal{H}} \subset \mathcal{H}$, and to solve the resulting (finitely constrained) programming problem. The design parameter vector $\widetilde{d}$, obtained by solving this problem, may not be useful for the original optimal multi-hazard design problem, because at some points of the resulting structure, belonging to $\mathcal{H} \setminus \widetilde{\mathcal{H}}$, the stresses may be too large.

In previous work [4, 5], we have proposed an iterative method for solving the optimal multi-hazard structural design problem (1), where each iteration requires the numerical solution of a (finitely constrained) optimization problem of the form,

$$\begin{aligned} &\min f(d) \\ \text{such that } &g(d) \leq 0\,, \\ &g^1(d,h) \leq -\varepsilon, \ \ \forall h \in \mathcal{H}_k\,, \\ &\ldots \\ &g^p(d,h) \leq -\varepsilon, \ \ \forall h \in \mathcal{H}_k\,, \end{aligned} \tag{2}$$

as well as the computation of the following quantities:

$$h^i_j = \text{argmax}_{h \in \mathcal{H}}\, g^i_j(\widetilde{d}, h), \quad i = 1, \ldots, p,\ j = 1, \ldots, m_i. \tag{3}$$

The approach from [4, 5] can be formalized as the following generic algorithm:

Algorithm 1

Select a (small) finite subset $\mathcal{H}_0 \subset \mathcal{H}$ and a tolerance $\varepsilon > 0$;

Set $k \leftarrow 0$;

repeat

Find the solution $\widetilde{d}$ of the optimization problem (2);

If $\widetilde{d}$ satisfies the constraints of (1),

then return $\widetilde{d}$ as the solution of the problem and **stop**;

Update $\mathcal{H}_k$ to obtain a larger subset $\mathcal{H}_{k+1} \subset \mathcal{H}$;

Set $k \leftarrow k + 1$.

continue

The updating of $\mathcal{H}_k$ in the above algorithm was done by adding to $\mathcal{H}_k$ the points at which the maximum violations of the constraints in (1) are attained and, possibly, some randomly generated points in $\mathcal{H}$. The addition of these randomly generated points had a stabilization effect on the algorithm. However, because of the randomness, the behavior of the algorithm varied slightly, both in the number of iterations before termination and the CPU time, each time it was used.

In the present paper we propose a purely deterministic variant of Algorithm 1, that turns out to be very stable, and outperforms most of time the random algorithms from [4,5]. For the sake of simplicity, we present the new algorithm only on the simple example considered in [4,5]. This example will be presented in Section 2. The description of our adaptive iterative strategy is contained in Section 3. We give some numerical results in Section 4, showing the robustness and efficiency of our approach.

2 A simple example of multi-hazard structural design

In [4, 5] we have examined a structure consisting of equally spaced steel columns supporting a long, straight horizontal metal pipe filled with water, and have considered the problem of designing a typical supporting column. The steel column has hollow elliptical cross sections with constant thickness w. Geometrically, the column is defined as the region between two frustums of cones with elliptical cross sections, and given hight H. The cross sections of the bottom and the top of the outer frustum are the ellipses that can be represented in the xy plane by the equations

$$\frac{x^2}{a^2} + \frac{y^2}{b^2} = 1 , \quad \frac{x^2}{A^2} + \frac{y^2}{B^2} = 1.$$

We assume that the two ellipses are similar, in the sense that $a/b = A/B$, which is equivalent to stating that they have the same eccentricity. Therefore A can be expressed as

$$A = \frac{aB}{b} . \tag{4}$$

Therefore, we have only four design variables a, b, B and w that have to satisfy the following bound constraints

$$w_{min} \leq w \leq a_{min} \leq a \leq b \leq B \leq B_{max} \, , \tag{5}$$

where the quantities $w_{min} < a_{min} < B_{max}$ are specified.

A vertical force denoted by V, equal to the weight of the pipe and the water it contains, is applied concentrically at the top of the column. Horizontal forces are applied at the center line of the pipe. They can be due to seismic effects or to wind effects. The seismic forces are denoted by F^e, can act in any direction, and are the same for all directions. The wind forces also act from any direction and are denoted by $F^w \cos\alpha$, where α is the angle between the wind direction and the normal to the longitudinal axis of the pipe. They are largest when normal to the pipe and are assumed to be negligible when parallel to the pipe.

It is easily shown that at coordinate h from the top of the column the compression stress due to gravity loading (defined as the sum of the vertical concentric load V and the self-weight of the column above the coordinate h, divided by the area of the cross section at coordinate h) is given by

$$s_g(a, b, B, w, h) = \frac{2bHV + 2bH\pi(a+b-w)w\gamma h + (a+b)(B-b)\pi w\gamma h^2}{2\pi(H(a+b-w)b + (a+b)(B-b)h)w} \, , \tag{6}$$

where γ is the specific weight of the column material.

It is also shown (see [5] for technical details) that the bending stress at coordinate h, induced by a force F that makes an angle α with the major axis of the ellipse and is applied at a point situated at distance c above the column, can be expressed by the formula

$$s_b(a, b, B, w, F, \alpha, h) = \frac{F(c+h)\, q_0(a(h), b(h), \alpha)}{\mu(a(h), b(h), w, \alpha)} \, , \tag{7}$$

where

$$a(h) = \left(1 - \frac{h}{H}\right) a + \frac{h}{H}\frac{aB}{b} \, , \quad b(h) = \left(1 - \frac{h}{H}\right) b + \frac{h}{H} B \, ,$$

$$q_0(a, b, \alpha) = \sqrt{.5(a^2 + b^2 + (b^2 - a^2)\cos(2\alpha))} \, ,$$

$$\mu(a, b, w, \alpha) = \frac{\pi}{8} ab \left(a^2 + b^2 + \left(b^2 - a^2\right)\cos(2\alpha)\right)$$

$$- \frac{\pi}{8}(a-w)(b-w)\left((a-w)^2 + (b-w)^2 + \left((b-w)^2 - (a-w)^2\right)\cos(2\alpha)\right) .$$

We assume that the wind force, of magnitude F^w, acts along the y-axis, i.e., along the major axis of the ellipse. The corresponding bending stress is obtained by taking $F = F^w$ and $\alpha = 0$ in (7),

$$s_b^w(a, b, B, w, h) = s_b(a, b, B, w, F^w, 0, h) \, . \tag{8}$$

Therefore the total stress in this case is

$$s^w(a, b, B, w, h) = s_g(a, b, B, w, h) + s_b^w(a, b, B, w, h)\,. \tag{9}$$

Since we would like this stress to be below the maximum admissible stress σ we obtain a constraint of the form

$$s^w(a, b, B, w, h) \le \sigma, \quad \forall j \in \mathcal{H}\,. \tag{10}$$

As the direction of the force F^e of the earthquake is not known, the corresponding constraint becomes

$$\begin{aligned} s^e(a, b, B, w, h) &= s_g(a, b, B, w, h) + \max_{\alpha \in [0,\pi]} s_b(a, b, B, w, F^e, \alpha, h) \le \sigma, \\ 0 &\le h \le H\,. \end{aligned} \tag{11}$$

The optimal design problem we are considering is to find design variables a, b, B, w satisfying the constraints (5), (10), and (11), such that the mass of the corresponding column is minimized. That mass is our objective function. It can be expressed in terms of the design variables as

$$\varphi(a, b, B, w) = \frac{H\pi\rho(a(b + B) + b(b + B - 2w))w}{2b}, \tag{12}$$

where ρ is the specific mass of the column material. We note that in the International System of Units (SI), the specific mass and the specific weight are related by the equation

$$\rho = \frac{\gamma}{9.80665}\,. \tag{13}$$

With the notation introduced above, our optimal multi-hazard structural design problem can be written as

$$\begin{aligned} &\min_{a,b,B,w} \ \varphi(a, b, B, w) \\ &\text{such that } s^w(a, b, B, w) \le \sigma\,, \\ &s^e(a, b, B, w) \le \sigma\,, \\ &w_{min} \le w \le a_{min} \le a \le b \le B \le B_{max}\,. \end{aligned} \tag{14}$$

This is a particular case of the general formulation (1), described in the introduction.

3 An iterative method

In this section we present our deterministic variant of Algorithm 1 for solving the semi-infinite programming problem (14). In this case we have $\mathcal{H} = [0, H]$, and we choose the initial finite subset of $\mathcal{H}$ to be

$$\mathcal{H}_0 = \left\{0\,,\ \frac{H}{3}\,,\ \frac{2H}{3}\,,\ H\right\}\,. \tag{15}$$

At a typical iteration we compute the solution $(\widetilde{a}, \widetilde{b}, \widetilde{B}, \widetilde{w})$ of the nonlinear optimization problem

$$\min_{a,b,B,w} \ \varphi(a, b, B, w) \tag{16}$$

$$\begin{aligned} \text{such that } s^{w}(a, b, B, w, h) &\leq \sigma - \frac{\varepsilon}{2\pi w}\,, \qquad h \in \mathcal{H}_k\,, \\ s^{e}(a, b, B, w, h) &\leq \sigma - \frac{\varepsilon}{2\pi w}\,, \qquad h \in \mathcal{H}_k\,, \\ w_{min} \leq w \leq a_{min} &\leq a \leq b \leq B \leq B_{max}\,. \end{aligned}$$

Then we compute the maximum stresses and denote

$$hW = \text{argmax}_{h\in[O,H]} s^{w}(\widetilde{a}, \widetilde{b}, \widetilde{B}, \widetilde{w}, h), \quad msW = s^{w}(\widetilde{a}, \widetilde{b}, \widetilde{B}, \widetilde{w}, hW). \tag{17}$$

$$hQ = \text{argmax}_{h\in[O,H]} \overline{s}^{\,e}(\widetilde{a}, \widetilde{b}, \widetilde{B}, \widetilde{w}, h), \quad msQ = s^{e}(\widetilde{a}, \widetilde{b}, \widetilde{B}, \widetilde{w}, hQ), \tag{18}$$

In our implementation the points hW and hQ are found within a tolerance of $.01\varepsilon$, by using a golden section method procedure [6, pp. 726].

If both msW and msQ are less than $\sigma - .1\varepsilon$ then we return $(\widetilde{a}, \widetilde{b}, \widetilde{B}, \widetilde{w})$ as the solution of the optimal multi-hazard structural design problem (14). Otherwise we update $\mathcal{H}_k$ and start a new iteration. It turns out that there are some advantages if at the end of the first iteration ($k = 0$)we construct $\mathcal{H}_1$ by enlarging the set

$$\left\{0\,,\ \frac{H}{2}\,,\ H\right\}\,, \tag{19}$$

rather than the set (15). For $k >$ we always construct $\mathcal{H}_{k+1}$ by enlarging $\mathcal{H}_k$. The precise updating method is described in the following algorithm:

Algorithm 2

Choose a tolerance $\varepsilon > 0$;

Consider the set $\mathcal{H}_0 \subset \mathcal{H}$ defined by (15);

Set $k \leftarrow 0$;

repeat

Compute the solution $\widetilde{d} = (\widetilde{a}, \widetilde{b}, \widetilde{B}, \widetilde{w})$ of (16);

Compute (17) and (18) within a tolerance of $.01\varepsilon$;

If $\max\{msW, msQ\} < \sigma - .1\varepsilon$,

then return $\widetilde{d}$ as the solution of the problem and **stop**;

If $k = 0$, then replace $\mathcal{H}_k$ with the set (19);

If $msW > \sigma - .1\varepsilon$, then set $\mathcal{H}_{k+1} = \mathcal{H}_k \bigcup\{hW\}$;

If $msQ > \sigma - .1\varepsilon$, then set $\mathcal{H}_{k+1} = \mathcal{H}_k \bigcup\{hQ\}$;

Set $k \leftarrow k + 1$.

continue

4 Numerical results

In the numerical results to be presented in this section we have taken the following values for the constants defining our model

$$H = 3.6\,m,\ c = 0.6\,m,\ V = 5000\,N,\ \gamma = 76820\,N/m^3,$$
$$F^e = 6000\,N,\ F^w = 8000\,N,\ \sigma = 165\,MPa,$$
$$w_{min} = .003\,m,\ a_{min} = .01\,m,\ B_{max} = .24\,m.$$

Previous numerical experiments reported in [5] showed that interior point methods (see [7] and the literature cited therein) are very efficient in solving nonlinear programming problems in structural optimization.

The numerical results from this section were obtained by using the *Mathematica* function `FindMinimum` with the option `Method -> "InteriorPoint"`, and the following trivial starting point

$$a = 0.4B_{max}, \quad b = 0.5B_{max}, \quad B = 0.9B_{max}, \quad w = 1.5w_{min}\,.$$

By using Algorithm 2 we obtained the following values for the design parameters:

$$a = 0.0532\,m,\ b = 0.0915\,m,\ B = 0.2031\,m,\ w = 0.003\,m.$$

The (optimal) weight of the corresponding column is 61.1 kg. Algorithm 2 needed 6 iterations to find this solution and the total CPU time (on a Dell Optiplex machine with 1 GB of RAM and a 3.20 GHz Pentium 4 processor) was 2.8 seconds. The same values for the design variables were obtained with the method described in [4,5]. As we mentioned before that method has a random component so that the CPU time varies each time the program is run. We run the program 10 times for the problem considered in this section. While the program gave the same solution in all of the runs the execution time varied. The average CPU time was 5.4 seconds with a standard deviation of one second. Also the average number of iterations was 8.5 with a standard deviation of almost .97 iterations.

5 Conclusions

In this paper we have described an iterative method that can be easily coupled with interior point methods in order to solve efficiently semi-infinite programming problems arising in multi-hazard structural design. Our method is robust and strictly deterministic. The new method appears to be more efficient, both in terms of number of iterations and computational time, than our previous algorithms which were not completely deterministic.

Disclaimer Certain commercial software products are identified in this paper in order to adequately specify the computational procedures. Such identification does not imply recommendation or endorsement by the National Institute of Standards and Technology nor does it imply that the software products identified are necessarily the best available for the purpose.

References

[1] Goberna, M.Á. & López, M.A., (eds.) *Semi-infinite programming*, volume 57 of *Nonconvex Optimization and its Applications*. Kluwer Academic Publishers: Dordrecht, pp. xviii+386, 2001. Recent advances, Papers from the conference (SIP'99) held in Alicante, September 1999.

[2] Stein, O., *Bi-level strategies in semi-infinite programming*, volume 71 of *Nonconvex Optimization and its Applications*. Kluwer Academic Publishers: Boston, MA, pp. xxviii+202, 2003.

[3] Weber, G.W., *Generalized semi-infinite optimization and related topics*, volume 29 of *Research and Exposition in Mathematics*. Heldermann Verlag: Lemgo, pp. xxii+361, 2003.

[4] Potra, F.A. & Simiu, E., Optimization and structural multi-hazard design. *Journal of Engineering Mechanics*, (submitted for publication), 2008.

[5] Potra, F.A. & Simiu, E., Multi-hazard design: A structural optimization approach. *Journal of Optimization Theory and Applications*, (submitted for publication), 2009.

[6] Bertsekas, D.P., *Nonlinear programming*. Athena Scientific, Belmont, MA, 2nd edition, 1999.

[7] Wright, M.H., The interior-point revolution in optimization: history, recent developments, and lasting consequences. *Bull Amer Math Soc (NS)*, **42(1)**, pp. 39–56 (electronic), 2005.

Mixed-integer non-linear programming approach to structural optimization

S. Kravanja
University of Maribor, Faculty of Civil Engineering, Maribor, Slovenia

Abstract

The paper presents the Mixed-Integer Non-Linear Programming (MINLP) approach to structural optimization. MINLP is a combined discrete/continuous optimization technique, where discrete binary 0-1 variables are defined for optimization of discrete alternatives and continuous variables for optimization of parameters. The MINLP optimization is performed through three steps: i.e. the generation of a mechanical superstructure, the modelling of an MINLP model formulation and the solution of the defined MINLP problem. As the discrete/continuous optimization problems are usually non-convex and highly non-linear, the Modified Outer-Approximation/Equality-Relaxation (OA/ER) algorithm is applied for the optimization. The accompanied Linked Multilevel Hierarchical Strategy (LMHS) is developed to accelerate the convergence of the mentioned algorithm. Two examples are presented at the end of the paper.
Keywords: mixed-integer non-linear programming, MINLP, structural optimization, structural synthesis.

1 Introduction

The paper presents the Mixed-Integer Non-Linear Programming (MINLP) approach to structural optimization. MINLP handles continuous and discrete binary 0-1 variables simultaneously. While continuous variables are defined for the continuous optimization of parameters (dimensions, stresses, strains, weights, costs, etc.), discrete variables are used to express discrete decisions, i.e. usually the existence or non-existence of structural elements inside the defined structure. Different materials, standard dimensions and rounded dimensions may also be defined as discrete alternatives. Since continuous and discrete optimizations are carried out simultaneously, the MINLP approach also finds optimal continuous

WIT Transactions on The Built Environment, Vol 106, © 2009 WIT Press
www.witpress.com, ISSN 1743-3509 (on-line)
doi:10.2495/OP090031

parameters, structural topology, material, standard and rounded dimensions simultaneously.

The MINLP discrete/continuous optimization problems of structural optimization are in most cases comprehensive, non-convex and highly non-linear. The MINLP optimization approach is proposed to be performed through three steps:

- *Generation of MINLP superstructure* for different structural/topology and other design alternatives that are candidates for a feasible and optimal solution.
- *Development of a special MINLP model formulation* for the defined superstructure in an equation oriented environment.
- *Solution of the defined MINLP model*, performed by suitable MINLP algorithm and strategies, which in the simultaneous MINLP optimization approach yield an optimal structural topology, material, shape, standard cross-sections and rounded dimensions.

Many different methods for solving MINLP problems have been developed in the near past. This paper reports the experience in solving the optimization problems by using the Outer-Approximation/Equality-Relaxation (OA/ER) algorithm [1, 2]. The Linked Multilevel Hierarchical strategy (LMHS) has been developed to accelerate the convergence of the mentioned algorithm. Since the number of discrete alternatives and defined binary 0-1 variables are usually too high for normal solution of the MINLP, a special reduction procedure is developed to reduce automatically the number of binary variables on a reasonable level.

Two examples are presented at the end of the paper. The first example shows the material and standard dimension optimization of a 32 m long composite I beam floor system and the second example deals with the simultaneous topology, material and standard section optimization of a single-storey industrial steel building structure.

2 Mechanical superstructure

The MINLP optimization approach to structural optimization requires the generation of an MINLP mechanical superstructure composed of various topology and design alternatives that are all candidates for a feasible and optimal solution. While topology alternatives represent different selections and interconnections of corresponding structural elements, design alternatives include different materials, standard and rounded dimensions.

The superstructure is typically described by means of unit representation: i.e. structural elements and their interconnection nodes. Each potential topology alternative is represented by a special number and a configuration of selected structural elements and their interconnections; each structural element may in addition have different material, standard and rounded dimension alternatives.

Therefore, the main goal is to find within the given superstructure a feasible structure that is optimal with respect to topology, material, standard and rounded dimensions as well as all defined continuous parameters.

3 MINLP model formulation for mechanical superstructures

It is assumed that a general non-convex and non-linear discrete/continuous optimization problem can be formulated as an MINLP problem (MINLP-G) in the form:

$$\begin{aligned} \min \quad & z = \boldsymbol{c}^T \boldsymbol{y} + f(\boldsymbol{x}) \\ \text{s.t.} \quad & \boldsymbol{h}(\boldsymbol{x}) = \boldsymbol{0} \\ & \boldsymbol{g}(\boldsymbol{x}) \le \boldsymbol{0} \\ & \boldsymbol{By} + \boldsymbol{Cx} \le \boldsymbol{b} \\ & \boldsymbol{x} \in X = \{\boldsymbol{x} \in R^n : \boldsymbol{x}^{\mathrm{LO}} \le \boldsymbol{x} \le \boldsymbol{x}^{\mathrm{UP}}\} \\ & \boldsymbol{y} \in Y = \{0,1\}^m \end{aligned} \qquad \textbf{(MINLP-G)}$$

where $\boldsymbol{x}$ is a vector of continuous variables specified in the compact set X and $\boldsymbol{y}$ is a vector of discrete, mostly binary 0-1 variables. Functions $f(\boldsymbol{x})$, $\boldsymbol{h}(\boldsymbol{x})$ and $\boldsymbol{g}(\boldsymbol{x})$ are non-linear functions involved in the objective function z, equality and inequality constraints, respectively. Finally, $\boldsymbol{By}+\boldsymbol{Cx}\le\boldsymbol{b}$ represents a subset of mixed linear equality/inequality constraints.

The above general MINLP model formulation has been adapted for the optimization of mechanical superstructures (MINLP-MS). The resulted formulation is more specific, particularly in variables and constraints. It can be used also for the modelling of any mechanical structure. It is given in the following form:

$$\begin{aligned} \min \quad & z = \boldsymbol{c}^T \boldsymbol{y} + f(\boldsymbol{x}) \\ \text{s.t.} \quad & \boldsymbol{h}(\boldsymbol{x}) = \boldsymbol{0} \\ & \boldsymbol{g}(\boldsymbol{x}) \le \boldsymbol{0} \\ & \boldsymbol{A}(\boldsymbol{x}) \le \boldsymbol{a} \\ & \boldsymbol{Ey} \le \boldsymbol{e} \\ & \boldsymbol{Dy}^e + \boldsymbol{R}(\boldsymbol{x}) \le \boldsymbol{r} \\ & \boldsymbol{Ky}^e + \boldsymbol{L}(\boldsymbol{d}^{cn}) \le \boldsymbol{k} \\ & \boldsymbol{Py} + \boldsymbol{S}(\boldsymbol{d}^m) \le \boldsymbol{s} \\ & \boldsymbol{My} + \boldsymbol{N}(\boldsymbol{d}^{st}) \le \boldsymbol{t} \\ & \boldsymbol{x} \in X = \{\boldsymbol{x} \in R^n : \boldsymbol{x}^{\mathrm{LO}} \le \boldsymbol{x} \le \boldsymbol{x}^{\mathrm{UP}}\} \\ & \boldsymbol{y} \in Y = \{0,1\}^m \end{aligned} \qquad \textbf{(MINLP-MS)}$$

The MINLP-MS model formulation for mechanical superstructures is proposed to be described as follows:

- Included are continuous variables $\boldsymbol{x}=\{\boldsymbol{d}, \boldsymbol{p}\}$ and discrete binary variables $\boldsymbol{y}=\{\boldsymbol{y}^e, \boldsymbol{y}^m, \boldsymbol{y}^{st}\}$. Continuous variables are partitioned into design variables $\boldsymbol{d}=\{\boldsymbol{d}^{cn}, \boldsymbol{d}^m, \boldsymbol{d}^{st}\}$ and into performance (nondesign) variables $\boldsymbol{p}$, where subvectors $\boldsymbol{d}^{cn}$,$\boldsymbol{d}^m$ and $\boldsymbol{d}^{st}$ stand for continuous dimensions, discrete materials and standard dimensions, respectively. Subvectors of binary variables $\boldsymbol{y}^e$, $\boldsymbol{y}^m$ and $\boldsymbol{y}^{st}$ denote the potential existence of structural elements inside the superstructure (the topology determination) as well as the potential selection of the discrete materials and standard dimension alternatives, respectively.
- The economical (or mass) objective function z involves fixed cost charges in the linear term $\boldsymbol{c}^{\mathrm{T}}\boldsymbol{y}$ and dimension dependant costs in the term $f(\boldsymbol{x})$.
- Parameter non-linear and linear constraints $\boldsymbol{h}(\boldsymbol{x})=\boldsymbol{0}$, $\boldsymbol{g}(\boldsymbol{x})\leq \boldsymbol{0}$ and $A(\boldsymbol{x})\leq \boldsymbol{a}$ represent the rigorous system of the design, loading, stress, deflection, stability, etc. constraints known from the structural analysis.
- Integer linear constraints $\boldsymbol{Ey}\leq \boldsymbol{e}$ are proposed to describe relations between binary variables.
- Mixed linear constraints $\boldsymbol{Dy}^e+\boldsymbol{R}(\boldsymbol{x})\leq \boldsymbol{r}$ restore interconnection relations between currently selected or existing structural elements (corresponding y^e = 1) and cancel relations for currently disappearing or nonexisting elements (corresponding $y^e = 0$).
- Mixed linear constraints $\boldsymbol{Ky}^e+\boldsymbol{L}(\boldsymbol{d}^{cn})\leq \boldsymbol{k}$ are proposed to define the continuous design variables for each existing structural element. The space is defined only when the corresponding structure element exists (y^e =1), otherwise it is empty.
- Mixed linear constraints $\boldsymbol{Py}+\boldsymbol{S}(\boldsymbol{d}^m)\leq \boldsymbol{s}$ define discrete materials $\boldsymbol{d}^m$. The disrete materials are defined in a similar way as the standard dimensions.
- Mixed linear constraints $\boldsymbol{My}+\boldsymbol{N}(\boldsymbol{d}^{st})\leq \boldsymbol{t}$ define standard discrete design variables $\boldsymbol{d}^{st}$. Each standard dimension d^{st} is determined as a scalar product between its vector of standard discrete dimension constants $\boldsymbol{q}$ and its vector of binary variables $\boldsymbol{y}^{st}$, eqn. (1). Only one discrete value can be selected for each standard dimension since the sum of binary variables must be equal one, eqn. (2):

$$d^{\mathrm{st}} = \sum_{i\in I} q_i y_i^{\mathrm{st}} \tag{1}$$

$$\sum_{i\in I} y_i^{\mathrm{st}} = 1 \tag{2}$$

The uper MINLP-MS model formulation for mechanical superstructures is presented in the condensed form. More about MINLP see Kravanja et al. [2–5].

4 Solving an MINLP problem

After the MINLP model formulation is developed, the defined MINLP optimization problem is solved by the use of suitable MINLP algorithm and strategies. A general MINLP class of optimization problems can be solved in principle by the following algorithms and their extensions:

- the *Nonlinear Branch and Bound*, NBB, proposed and used by many authors, e.g. Beale [6], Gupta and Ravindran [7];
- the *Sequential Linear Discrete Programming* method, SLDP, by Olsen and Vanderplaats [8] and Bremicker et al. [9];
- the *Extended Cutting Plane* method by Westerlund et al. [10];
- the *Generalized Benders Decomposition*, GBD, by Benders [11], Geoffrion [12];
- the *Outer-Approximation /Equality Relaxation algorithm*, OA/ER, by Kocis and Grossmann [13];
- the *Feasibility Technique* by Mawengkang and Murtagh [14]; and
- the *LP/NLP based Branch and Bound* algorithm by Quesada and Grossmann [15].

4.1 Modified OA/ER algorithm

The OA/ER algorithm seems to be one of the most efficient algorithm to solve large-scale MINLP problems, when NLP subproblems are expensive and difficult to solve. The OA/ER algorithm consists of solving an alternative sequence of Non-linear Programming (NLP) optimization subproblems and Mixed-Integer Linear Programming (MILP) master problems. The former corresponds to continuous optimization of parameters for a mechanical structure with fixed topology (and material, standard and rounded dimensions) and yields an upper bound to the objective to be minimized. The latter involves a global approximation to the superstructure of alternatives in which new topology, materials, standard and rounded dimensions are identified so that its lower bound does not exceed the current best upper bound. The search of a convex problem is terminated when the predicted lower bound exceeds the upper bound, otherwise it is terminated when the NLP solution can be improved no more. The convergence is usually achieved in a few MINLP iterations. The OA/ER algorithm guarantees the global optimality of solutions for convex and quasi-convex optimization problems.

The OA/ER algorithm as well as all other MINLP algorithms do not generally guarantee that the solution found is the global optimum. This is due to the presence of non-convex functions in the models that may cut off the global optimum. In order to reduce undesirable effects of nonconvexities, the Modified OA/ER algorithm was proposed by Kravanja and Grossmann [1] by which the following modifications are applied for the master problem: deactivation of linearizations, decomposition and deactivation of the objective function linearization, use of the penalty function, use of the upper bound on the objective function to be minimized as well as a global convexity test and a validation of the outer approximations.

4.2 Linked multilevel hierarchical strategy, LMHS

The optimal solution of comprehensive non-convex and non-linear MINLP problem with a high number of discrete decisions is in general very difficult to be obtained. For this purpose, the LMHS strategy has been developed to accelerate the convergence of the OA/ER algorithm. Using the LMHS strategy, we decompose the original integer space and original MINLP problem in a hierarchical manner into several subspaces and corresponding MINLP levels. Each time the next MINLP optimization level is performed, the current integer subspace is extended by the next integer subspace and prescreened, while the discrete decisions belonging to all of the remaining subspaces are approximated by the relaxed 0-1 variables. The levels are linked by accumulating outer-approximations and yield lower bounds to their next level objective functions to be minimized, which considerably improve the efficiency of the search. Decision levels are hierarchically classified as:

- The level of discrete topology and material alternatives (the highest level).
- The level of discrete standard dimension decisions (the middle level).
- The level of rounded continuous dimension decisions (the lower level).

Higher levels give lower bounds to the original objective function to be minimized while lower levels give upper bounds. The MINLP subproblems are iterated about each level until there are no improvements in the NLP solution. Thus, we start with the discrete topology and material optimization at the relaxed standard dimensions. When the optimal topology and materials are reached, we proceed with simultaneous discrete topology, material and standard dimension optimization at the second level. Finally, after the optimal topology, materials and standard dimensions are obtained, the MINLP is carried out once more for complete discrete decisions at the third level.

The optimization model may contain up to some ten thousands of binary 0-1 variables of alternatives. Most of them are subjected to standard and rounded dimensions. Since this number of 0-1 variables is too high for a normal solution of the MINLP, we developed a reduction procedure, which automatically reduces binary variables for standard and rounded dimension alternatives into a reasonable number. In the optimization at the second and third level are included only those 0-1 variables which determine standard and rounded dimension alternatives close to continuous dimensions, obtained at previous MINLP optimization iterations.

It should be noted, that the LMHS strategy can solve convex problems to global optimal solutions.

5 Numerical examples

The MINLP optimization approach is illustrated by two examples. The first example shows the material and standard dimension optimization of a 32 m long composite I beam floor system and the second one deals with the simultaneous topology, material and standard sizing optimization of an industrial steel building structure.

MINLP optimization models for the mentioned two different structures were developed. As an interface for mathematical modelling and data inputs/outputs GAMS (General Algebraic Modelling System) by Brooke et al. [16], a high level language, was used. The optimizations were carried out by a user-friendly version of the MINLP computer package MIPSYN, the successor of programs PROSYN by Kravanja and Grossmann [1] and TOP by Kravanja et al. [17]. The Modified OA/ER algorithm and the LMHS strategy were applied, where GAMS/CONOPT2 (Generalized reduced-gradient method), see Drud [18], was used to solve NLP subproblems and GAMS/Cplex 7.0 (Branch and Bound) [19] was used to solve MILP master problems. The structures were modelled in accordance with Eurocodes: Eurocode 4 [20] for the composite structure and Eurocode 3 [21] for steel building structure.

5.1 Material and standard dimension optimization of a composite I beam floor system

The first example presents the material and standard dimension optimization of a composite I beam floor system with the span of 32 m, subjected to the self-weight and to the uniformly distributed imposed load of 7.5 kN/m^2.

Table 1: Material and labour costs.

Material costs for structural steel S 235-S 355	1.0-1.2	EUR/kg
Material costs for reinforcing steel S 400	1.2	EUR/kg
Material costs for concrete C 25/30-C 50/60	100.0-120.0	EUR/m^3
Sheet-iron cutting costs	6.0	EUR/m^1
Cylindrical shear stud	1.2	EUR/piece
Welding costs	8.0	EURm^1
Anti-corrosion resistant painting costs (R30)	25.0	EUR/m^2
Panelling costs	10.0	EUR/m^2

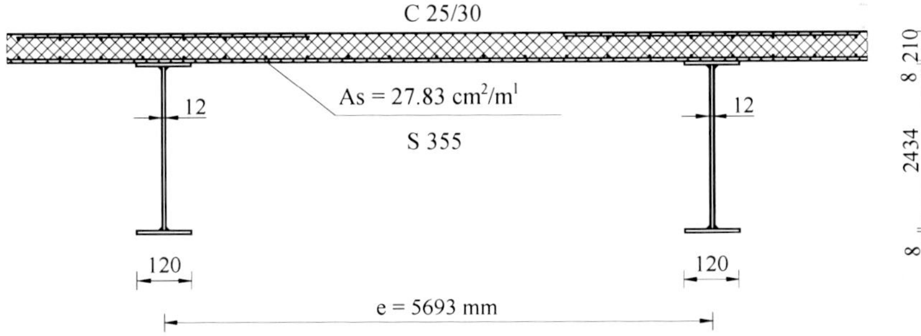

Figure 1: Optimal cross-section of the composite I-beam floor system.

The task of the MINLP optimization was to find the minimal manufacturing costs. The material and labour costs for the composite beams were thus accounted for in the economical type of the objective function, subjected to the given design, material, resistance and deflection constraints, defined according to Eurocodes 4 [20]. The material and labour costs for the composite beams considered are shown in Table 1. The superstructure comprised 6 different concrete strengths (C25, C30, C35, C40, C45, C50), 3 different structural steel grades (S 235, S 275, S 355), 48 various standard reinforcing steel sections as well as 9 different standard thickness of sheet-iron plates (from 8 mm to 40 mm) for webs and flanges separately.

The optimal result of 116.90 EUR/m^2 was obtained in the 2nd MINLP iteration, see Figure 1. Beside the optimal self-manufacturing costs, the optimal concrete strength C25/30, steel grade S 355, standard reinforcing wire mesh and the optimal standard thickness of webs and flanges were obtained.

5.2 Simultaneous topology, material and standard section optimization of a single-storey industrial steel building structure

The second example presents the topology, material and standard section optimization of a single-storey industrial steel building structure. The structure was consisted from equal non-sway steel portal frames, which are mutually connected with purlins and rails. The building was 27.5 meters wide, 72.5 meters long and 8.25 meters high. The mass of the roof was 0.20 kg/m^2 and of the facade cladding 0.15 kg/m^2. The structure was subjected to the self-weight and to the variable load. The variable imposed loads 2.00 kN/m^2 (snow) and 0.50 kN/m^2 (horizontal wind) were defined in the model input data.

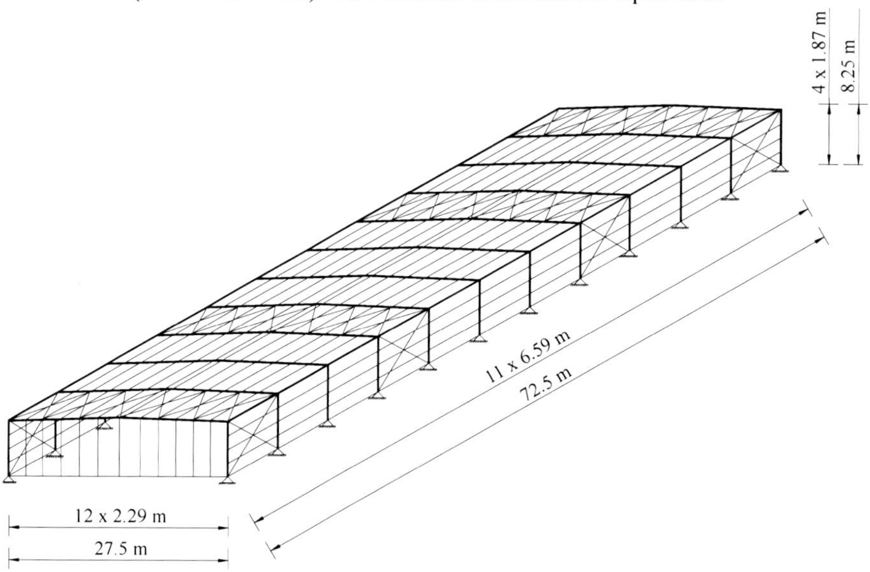

Figure 2: Optimal design of the single-storey industrial building.

The task of the performed MINLP optimization was to find the minimal structure mass. The mass objective function was thus defined for the optimization. The internal forces and deflections were calculated by the elastic first-order analysis. The design/dimensioning constraints were defined in accordance with Eurocode 3 [21]. The structure was checked for both the ultimate and serviceability limit states. The superstructure was generated in which all possible building's structures were embedded by 70 portal frame alternatives, 40 purlin alternatives, 20 rail alternatives, 3 discrete steel material alternatives and 24 different alternatives of standard hot rolled HEA sections (from HEA 100 to 1000) for each column, beam, purlin and rail separately.

The final optimal solution of 158.31 tons was obtained in the 4th MINLP iteration. The optimal solution represents the mentioned "minimal" structure mass as well as the building topology of 12 portal frames, 14 purlins and 12 rails, see Figure 2. Columns and beams are both designed from HEA 700, purlins from HEA 160 and rails from HEA 100 standard sections, see Figure 3.

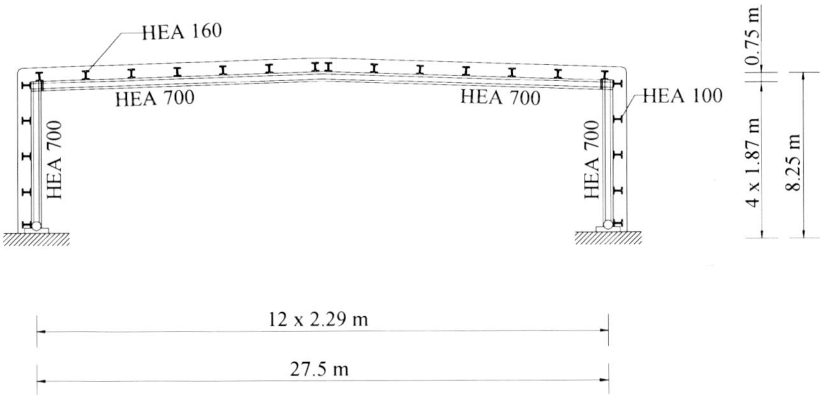

Figure 3: Optimal design of the portal frame.

6 Conclusions

The paper presents the Mixed-Integer Non-Linear Programming (MINLP) approach to structural optimization. The MINLP was found to be a successful optimization technique for solving structural optimization problems.

References

[1] Kravanja, Z. & Grossmann, I.E., New Developments and Capabilities in PROSYN - An Automated Topology and Parameter Process Synthesizer. *Computers chem. Eng.*, **18**, pp. 1097-1114, 1994.

[2] Kravanja, S., Kravanja, Z., Bedenik, B.S., The MINLP optimization approach to structural synthesis. Part I: A general view on simultaneous topology and parameter optimization. *Int. J. Numer. Methods Eng.*, **43**, pp. 263-292, 1998.

[3] Kravanja, S., Kravanja, Z., Bedenik, B.S., The MINLP optimization approach to structural synthesis. Part II: Simultaneous topology, parameter and standard dimension optimization by the use of the Linked two-phase MINLP strategy. *Int. J. Numer. Methods Eng.*, **43**, pp. 293-328, 1998.

[4] Kravanja, S., Kravanja, Z., Bedenik, B.S., The MINLP optimization approach to structural synthesis. Part III: Synthesis of roller and sliding hydraulic steel gate structures. *Int. J. Numer. Methods Eng.*, **43**, pp. 329-364, 1998.

[5] Kravanja, S., Šilih S., Kravanja, Z., The multilevel MINLP optimization approach to structural synthesis: the simultaneous topology, material, standard and rounded dimension optimization. *Internatinal Journal on Advances in Engineering Software*, **36**, pp. 568-583, 2005.
[6] Beale, E.M.L., *Integer Programming. The State of the Art in Numerical Analysis*, ed D. Jacobs, Academic Press, London, pp. 409-448, 1977.
[7] Gupta, O.K. & Ravindran, A., Branch and bound experiments in convex nonlinear integer programming. *Management Science*, **31**(12), pp.1533–1546, 1985.
[8] Olsen, G.R. & Vanderplaats, G.N., Method for Nonlinear Optimization with Discrete Design Variables. *AIAA Journal*, **27**(11), pp. 1584-1589, 1989.
[9] Bremicker, M., Papalambros, P.Y., Loh, H.T., Solution of Mixed-Discrete Structural Optimization Problems with a New Sequental Linearization Method. *Computers & Structures*, **37**(4), pp. 451-461, 1990.
[10] Westerlund, T., Pettersson, F., Grossmann, I.E., Optimization of pump configurations as a MINLP problem. *Computers chem. Engng.*, **18**(9), pp. 845-858, 1994.
[11] Benders, J.F., Partitioning Procedures for Solving Mixed-variables Programming Problems. *Numerische Mathematik*, **4**, pp. 238-252, 1962.
[12] Geoffrion, A.M., Generalized Benders Decomposition. *Journal of Optimization Theory and Applications*, **10**(4), pp. 237-260, 1972.
[13] Kocis, G.R. & Grossmann, I.E., Relaxation Strategy for the Structural Optimization of Process Flowsheets. *Ind. Engng Chem. Res.*, **26**, pp. 1869-1880, 1987.
[14] Mawengkang, H. & Murtagh, B.A., Solving Nonlinear Integer Programs with Large-Scale Optimization Software. *Annals of Operations Research*, **5**, pp. 425-437, 1986.
[15] Quesada, I. & Grossmann, I.E., An LP/NLP Based Branch and Bound Algorithm for Convex MINLP Optimization Problems. *Computers Chem. Engng.*, **16**, pp. 937-947, 1992.
[16] Brooke, A., Kendrick, D., Meeraus, A., *GAMS - A User's Guide*, Scientific Press, Redwood City, CA, 1988.
[17] Kravanja, S., Kravanja, Z., Bedenik, B.S., Faith, S., Simultaneous Topology and Parameter Optimization of Mechanical Structures, *Proceedings of the First European Conference on Numerical Methods in Engineering* (ed Ch. Hirsch et al.), Brussels, Belgium, Elsevier, Amsterdam, pp. 487-495, 1992.
[18] Drudd, A.S., CONOPT – A Large-Scale GRG Code. *ORSA Journal on Computing*, **6** (2), pp. 207-216, 1994.
[19] CPLEX User Notes, ILOG inc.
[20] Eurocodes 4, Design of composite structures, European Comitee for Standardization, 1992.
[21] Eurocode 3, Design of steel structures, European Committee for Standardization, 1992.

Section 2
Shape optimisation

A practical solution to the shape optimization problem of solid structures

M. Shimoda[1], S. Motora[2] & H. Azegami[3]
[1]*Department of Mechanical Design Engineering, Shonan Institute of Technology, Japan*
[2]*Graduate School of Shonan Institute of Technology, Japan*
[3]*Nagoya University, Japan*

Abstract

This paper presents a practical optimization method for the shape design of solid structures or 3-dimensional structures in order to obtain the optimal free boundary shape without any parameterization of the shape for optimization. A solution to the rigidity design problem of a solid structure under the assumption that the Neumann boundary is allowed to vary is presented. The compliance is minimized subject to a volume constraint and the state equation. Surface tractions, body forces and hydrostatic pressure are applied on the specified regions. This design problem is formulated as a non-parametric shape optimization problem. The shape gradient function is theoretically derived using the Lagrange multiplier method, the material derivative method and the adjoint variable method. With the shape gradient function and the traction method that was proposed by the authors as a gradient method in a Hilbert space, the smooth optimal shape can be easily obtained. This solution is applied to four design problems. The results obtained verified the effectiveness and practical utility of the proposed method for the shape design of solid structures with variable Neumann boundaries.

Keywords: solid structure, shape optimization, traction method, optimal shape, non-parametric optimization, adjoint variable, material derivative.

1 Introduction

Solid, 3-dimensional structures are widely used in the mechanical and structural components of human-made objects such as vehicles, electrical appliances and

WIT Transactions on The Built Environment, Vol 106, © 2009 WIT Press
www.witpress.com, ISSN 1743-3509 (on-line)
doi:10.2495/OP090041

architectures. Determining the optimal shape of components is a universal problem in all areas of mechanical and structural design. Product performance requirements for weight, cost, rigidity, strength, vibration and other attributes are becoming increasingly severe because of issues related to the depletion of natural resources and environmental protection. The development of shape optimization techniques is essential in order to obtain solutions efficiently and economically, as well as to overcome the limitations of current designs. In the shape design of solid components in particular, the optimal free-form design is required in order to fulfill the design requirements because solid structures have many design degrees of freedom. We have developed a non-parametric shape optimization method, which we call the "traction method" [1], and applied it to various shape design problems of 2D and 3D continua including plate and shell structures in our previous studies [2-3]. In optimizing a solid structure using a parametric shape optimisation method like the basis vector method, which is one of the best shape optimization techniques, we often encounter the problem of how to parameterize the shape or the problem of how to prepare the basis vectors. How many and which basis vectors are the best for obtaining the optimal shape? The result obtained also strongly depends on the basis vectors used. In contrast, with the traction method, we can easily obtain the smooth optimal boundary shape without any parameterization of the shape.

In this study, we applied the traction method to the shape optimization of solid structures under the assumption that the Neumann boundary was allowed to vary. It is important for structural designers to optimize the Neumann boundary for reducing the total amount of applied loads. The traction method is a gradient method in a Hilbert space. This study considered the rigidity design problem of a solid structure with variable Neumann boundaries. Surface tractions, body forces and hydrostatic pressure were applied on the specified regions. The compliance was minimized subject to a volume constraint and the state equation. The sensitivity function, i.e., the shape gradient function, and the optimality conditions for this problem were derived using the Lagrange multiplier method, the material derivative method and the adjoint variable method. In the traction method, the negative shape gradient function is applied in the normal direction to the design surface as an external force to vary the shape. This method was applied to four design problems involving a beam, a dam and a tower. The effect of each term in the shape gradient function on the optimum shape was also evaluated. The validity and practical utility of this method for the optimum shape design of solid structures with variable Neumann boundaries were verified by the results obtained.

2 Domain variation and material derivative for optimization

A technique for representing domain variation using the speed method [4] will be introduced briefly before formulating the shape optimization problem. A detailed explanation of this technique may be found in references [1] and [5].

As shown in Fig. 1, it is assumed that a linear elastic body having an initial domain of $\Omega \subset \mathbb{R}^3$ and boundary of $\Gamma \equiv \partial\Omega$ undergoes variation (i.e., the design velocity field) $\boldsymbol{V}$ such that its domain and boundary become Ω_s and Γ_s. The notation $\mathbb{R}$ indicates a set of positive real numbers. The domain variation can be expressed by a one-to-one mapping $T_S(\boldsymbol{X}): X \in \Omega \mapsto x \in \Omega_S$, $0 \le s < \varepsilon$. The notation s and ε indicate the iteration history of domain variation and a small positive number, respectively. Assuming a constraint is acting on the variation in the domain $\Theta \subset \Omega$, the infinitesimal variation of the domain can be given by

$$T_{s+\Delta s}(\boldsymbol{X}) = T_s(\boldsymbol{X}) + \Delta s \boldsymbol{V} , \tag{1}$$

where the design velocity field $\boldsymbol{V}$ is given as a derivative of $T_s(\boldsymbol{X})$ with respect to s and can be defined as a continuous function as

$$\boldsymbol{V}(\boldsymbol{x}) = \frac{\partial T_s}{\partial s}(T_s^{-1}(\boldsymbol{x})), \quad \boldsymbol{x} \in \Omega_s , \tag{2}$$

$$\boldsymbol{V} \in C_\Theta = \{\boldsymbol{V} \in C^1(\Omega;\mathbb{R}^3) \,|\, \boldsymbol{V} = \boldsymbol{0} \text{ in } \Theta\} . \tag{3}$$

The derivative of a response functional is obtained as indicated below. When a response functional J is given as a domain integral of a distributed function ϕ_s,

$$J = \int_{\Omega_s} \phi_s d\Omega . \tag{4}$$

The material derivative with respect to s $\dot{J}$ is given by the following expression.

$$\dot{J} = \int_{\Omega_s} \phi_s' d\Omega + \int_{\Gamma_s} \phi_s V_n d\Gamma , \tag{5}$$

where $V_n = n_i V_i$. The vector $\boldsymbol{n}$ is an outward unit normal vector. The notation $(\cdot)'$ indicates a shape (Lagrange) derivative with respect to s [4]. When a response functional J is given as a boundary integral of a distributed function ϕ_s,

$$J = \int_{\Gamma_s} \phi_s d\Gamma , \tag{6}$$

the material derivative $\dot{J}$ is given by

$$\dot{J} = \int_{\Gamma_s} \{\phi_s' + (\phi_{s,i} n_i + \phi_s \kappa) V_n\} d\Gamma , \tag{7}$$

where κ expresses the mean curvature when it is three dimensional.

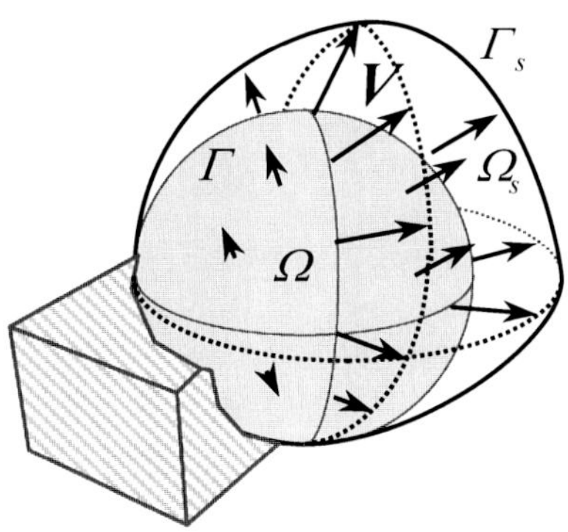

Figure 1: Domain variation by $\boldsymbol{V}$.

The tensor notation employed in this paper uses Einstein's summation convention and a partial differential notation $(\cdot)_{,i} = \partial(\cdot)/\partial x_i$.

3 Formulation of shape optimization problem of solid structure

The non-parametric shape optimization problem for the rigidity design of a solid structure with linear elastic material is solved. Consider body forces per unit volume $\boldsymbol{f}(\boldsymbol{x})$, surface forces per unit area $\boldsymbol{P}(\boldsymbol{x})$ and pressure $p(\boldsymbol{x})\boldsymbol{n}$ act on Ω, Γ_1 and Γ_2, respectively. Letting $l(\boldsymbol{v})$ denote the compliance as an index of rigidity, the rigidity design problem or the compliance minimization problem subject to constraints of volume and the state equation can be formulated as shown below.

$$\text{Find} \quad \Omega_s \ (\text{or } \boldsymbol{V}), \tag{8}$$

$$\text{that minimize} \quad l(\boldsymbol{v}), \tag{9}$$

$$\text{subject to} \quad a(\boldsymbol{v}, \boldsymbol{w}) = l(\boldsymbol{w}), \quad \forall \boldsymbol{w} \in U, \tag{10}$$

$$M \left(= \int_{\Omega_s} d\Omega\right) \le M_0 , \tag{11}$$

where M and M_0 denote the volume and its constraint value. The bilinear form $a(\boldsymbol{v}, \boldsymbol{w})$ that gives the variational strain energy and the linear form $l(\boldsymbol{w})$ that gives the variational potential energy due to the external force are defined as

$$a(\boldsymbol{v}, \boldsymbol{w}) = \int_{\Omega} e_{ijkl} v_{k,l} w_{i,j} d\Omega , \tag{12}$$

$$l(\boldsymbol{w}) = \int_{\Omega} f_i w_i d\Omega + \int_{\Gamma_1} P_i w_i d\Gamma + \int_{\Gamma_2} p(\boldsymbol{x}) n_i w_i d\Gamma , \tag{13}$$

where $\boldsymbol{v}$, $\boldsymbol{w}$ and e_{ijkl} are the displacement vector, the variational displacement vector and the elastic coefficients, respectively, and U denotes the suitably smooth function space that satisfies the displacement constraint condition.

Letting $\boldsymbol{w}$ and Λ denote the Lagrange multipliers for the state equation and the volume constraint, respectively, the Lagrangian functional L for this problem can be expressed as

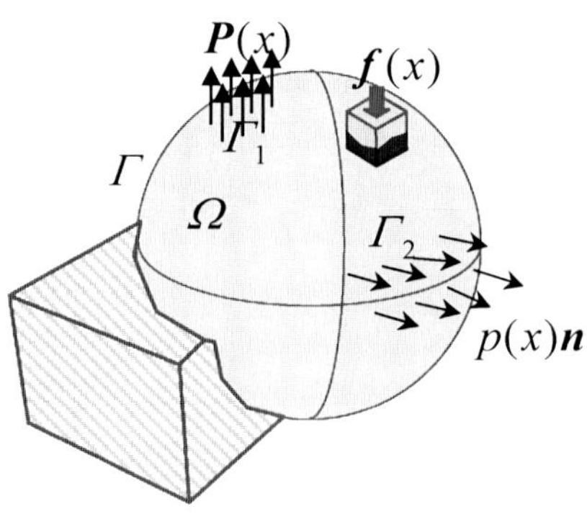

Figure 2: Boundary condition.

$$L(\Omega,\boldsymbol{v},\boldsymbol{w},\Lambda)=l(\boldsymbol{v})-a(\boldsymbol{v},\boldsymbol{w})+l(\boldsymbol{w})+\Lambda(M-M_0)\,. \tag{14}$$

Assuming that the body forces and the surface forces are constant with respect to s within the space ($\boldsymbol{f}'=\boldsymbol{P}'=0$) and that the material is homogeneous and constant ($\dot{e}_{ijkl}=e'_{ijkl}=0$), the material derivative $\dot{L}$ with respect to the domain variation of the Lagrangian functional L is expressed using the design velocity field $\boldsymbol{V}$ as follows:

$$\dot{L}=l(\boldsymbol{v}')-a(\boldsymbol{v}',\boldsymbol{w})-a(\boldsymbol{v},\boldsymbol{w}')+l(\boldsymbol{w}')+\Lambda'(M-M_0)+l_G(\boldsymbol{V})\,, \tag{15}$$

$$\begin{aligned} l_G(\boldsymbol{V}) &= \int_\Gamma \{f_i(v_i+w_i)-e_{ijkl}v_{k,l}w_{i,j}+\Lambda\}V_n d\Gamma \\ &+\int_{\Gamma_1}[\{P_{i,j}n_j(v_i+w_i)+P_i(v_{i,j}+w_{i,j})n_j+P_i(v_i+w_i)\kappa\}V_n]d\Gamma \\ &+\int_{\Gamma_2}[p'(v_i+w_i)n_i+\mathrm{div}(pv_i+pw_i)V_n]d\Gamma\;. \end{aligned} \tag{16}$$

The optimality conditions of this functional L with respect to $\boldsymbol{v}$, $\boldsymbol{w}$ and Λ are expressed as shown below.

$$a(\boldsymbol{v},\boldsymbol{w}')=l(\boldsymbol{w}'),\ \forall \boldsymbol{w}'\in U\ , \tag{17}$$

$$a(\boldsymbol{v}',\boldsymbol{w})=l(\boldsymbol{v}'),\ \forall \boldsymbol{v}'\in U\ , \tag{18}$$

$$\Lambda(M-M_0)=0,\ \ M-M_0=0,\ \ \Lambda\geq 0\,, \qquad (19)(20)(21)$$

where eqn. (17) is the governing equation of $\boldsymbol{v}$ which coincides with the state eqn. (10), and eqn. (18) is the governing equation of $\boldsymbol{w}$. The Lagrange multiplier Λ is determined so as to satisfy eqns. (19)–(21). Further, a comparison of governing eqn. (17) and adjoint eqn. (18) yields the following self-adjoint relationship:

$$\boldsymbol{v}=\boldsymbol{w}\,. \tag{22}$$

By substituting $\boldsymbol{v}$ (or $\boldsymbol{w}$) into eqn. (15), the material derivative $\dot{L}$ can be expressed as the dot product of the shape sensitivity function (i.e., shape gradient function) $\boldsymbol{G}$ and the design velocity field $\boldsymbol{V}$ as shown in eqns. (23)–(26).

$$\begin{aligned} \dot{L}=l_G(\boldsymbol{V}) &\equiv \int_\Gamma G\boldsymbol{n}\cdot\boldsymbol{V}d\Gamma \equiv \int_\Gamma \boldsymbol{G}\cdot\boldsymbol{V}d\Gamma \\ &= \int_\Gamma \boldsymbol{G}\cdot\boldsymbol{V}d\Gamma+\int_{\Gamma_1}\boldsymbol{G}_1\cdot\boldsymbol{V}d\Gamma+\int_{\Gamma_2}\boldsymbol{G}_2\cdot\boldsymbol{V}d\Gamma\ , \end{aligned} \tag{23}$$

$$\boldsymbol{G}=\{f_i(v_i+w_i)-e_{ijkl}v_{k,l}w_{i,j}+\Lambda\}\boldsymbol{n}\ , \tag{24}$$

$$\boldsymbol{G}_1=\{P_{i,j}n_j(v_i+w_i)+P_i(v_{i,j}+w_{i,j})n_j+P_i(v_i+w_i)\kappa\}\boldsymbol{n}\,, \tag{25}$$

$$\boldsymbol{G}_2=\{p'(v_i+w_i)n_i+\mathrm{div}(pv_i+pw_i)\}\boldsymbol{n}\ . \tag{26}$$

Since the shape gradient function has been derived, the traction method can be applied.

4 Traction method and shape optimization system

The traction method [1] is a gradient method in a Hilbert space. In the traction method, the negative shape gradient function $-\boldsymbol{G}$ is applied in the normal direction to the design boundary as an external traction force, i.e., as a Neumann condition to vary the shape. We call this process the velocity analysis. The resultant displacement field (i.e., design velocity field) $\Delta s\boldsymbol{V}$ represents the amount of domain variation added to the original shape to update it. Using this method, the smooth domain variation that minimizes the objective functional can be obtained. By repeating the stress analysis and the adjoint analysis that yield the shape gradient function, the velocity analysis and the updating of the shape by $\Delta s V$, the optimum shape can be obtained. The process of the optimization system based on this method is schematized in Fig. 3. Other advantages of this method are summarized as follows: (1) it is not necessary to parameterize the shape unlike the basis vector method, because all nodes on the design domain can be moved as the design variable, (2) it is not necessary to refine the mesh, because the entire domain can be mapped by the traction force, (3) it assures smooth boundary shapes without any zigzagging, because the elastic tensor serves as a smoother, (4) it can be easily implemented in combination with a commercial FEM code, which means it has generality and practical utility for actual design work. More details of the traction method involving the verification of smoothness are given in reference [6].

The governing equation of the velocity analysis with the Neumann condition is given as

$$a(\boldsymbol{V},\boldsymbol{w}) = -l_G(\boldsymbol{w}), \quad \forall \boldsymbol{w} \in C_\Theta \ . \tag{27}$$

Equation (27) can be solved by a standard finite element analysis.

The mean curvature κ in eqn. (25) for the C_0-continuity surface on the linear solid elements of a FE model was approximated by differentiating the Bezier surface obtained by the method of least squares. For a quadratic solid element, it can be calculated by using its shape function.

It can be confirmed that the domain variation $\boldsymbol{V}$ determined by the velocity analysis reduces the Lagrangian functional L. When the state equation, the adjoint equation and the constraints are satisfied, the perturbation expansion of L can be written as

$$\Delta L = l_G(\Delta s \boldsymbol{V}) + O(|\Delta s|) . \tag{28}$$

Substituting eqn. (27) into eqn. (28) and taking into account the positive definitiveness of $a(\boldsymbol{v},\boldsymbol{w})$, based on the positive definitiveness of the elastic tensor e_{ijkl}

$$\exists \alpha > 0 : a(\xi,\xi) \geq \alpha \|\xi\|^2 , \ \forall \xi \in U , \tag{29}$$

the following relationship is obtained when Δs is sufficiently small:

$$\Delta L = -a(\Delta s \boldsymbol{V}, \Delta s \boldsymbol{V}) < 0 \ . \tag{30}$$

This relationship definitely reduces the Lagrangian functional in the process of changing the domain using the velocity field $\boldsymbol{V}$ determined by eqn. (27). By repeating the stress analysis for evaluating the shape gradient function, the velocity analysis and the updating of the shape by $\Delta s \boldsymbol{V}$, the objective functional is minimized, resulting in the smooth optimum shape.

5 Results of shape optimization

To confirm the validity of the proposed solution, it was applied to four design problems. In all four examples, it was assumed that the body forces, the surface

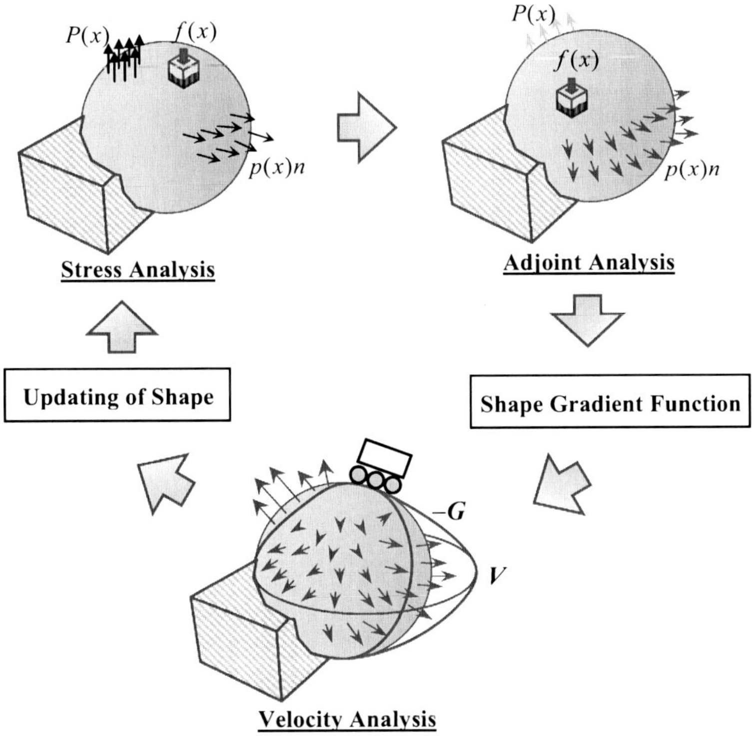

Figure 3: Schematic diagram of traction method.

forces and the pressure were constant with respect to s within the space ($\boldsymbol{f}' = \boldsymbol{P}' = p'(\boldsymbol{x})\boldsymbol{n} = 0$).

5.1 Short span beam problem

A short span beam fixed at both ends was optimized. This short span beam is subjected to both shearing stress and bending stress. The boundary conditions of the stress analysis and of the velocity analysis are shown in Fig. 4(a) and (b), respectively. In the stress analysis, downward uniformly distributed loads P ($P_{i,j} = 0$, or $\dot{P} = 0$) were applied on the upper and lower variable Neumann boundaries, where the design boundaries to be optimized were located. In the velocity analysis, the length and width were kept constant. A constant volume constraint was applied. The initial shape and the optimal shape that was calculated by using eqn. (24) (except the term for body force $f_i(v_i + w_i)$) and eqn. (25) as the shape gradient function, are shown respectively in Fig. 5(a) and (c). For comparison, the final shape calculated by only using the strain energy density (SED) and Λ in eqn. (24) as the shape gradient function is shown in Fig. 5(b), since SED is often used in place of the shape gradient function as a convenient index for the rigidity design. It is seen that both shapes (b) and (c) obtained have smooth boundaries. Figure 6 shows iteration histories of the volume and compliance in the optimization process. The values are normalized to the values of the initial shape. Both compliances are minimized while satisfying the volume constraint, but the rate of reduction is different; the rate for (b) is 17% while that for (c) is 19%. It is also seen that the optimal shape (c) is clearly different from the shape (b) at the beam center. These results confirm the effectiveness of the derived shape gradient function and our solution.

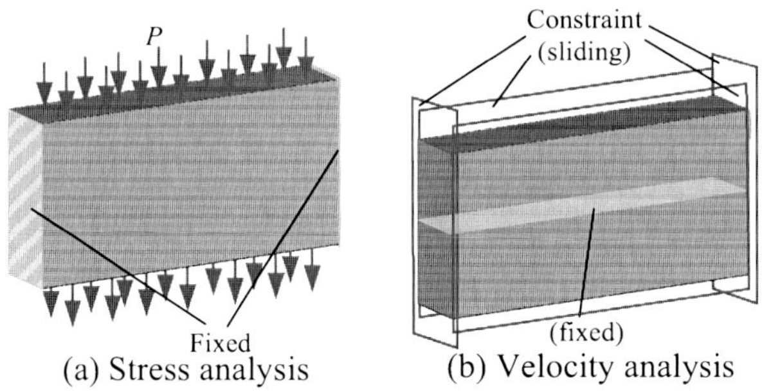

Figure 4: Boundary conditions for short span beam problem.

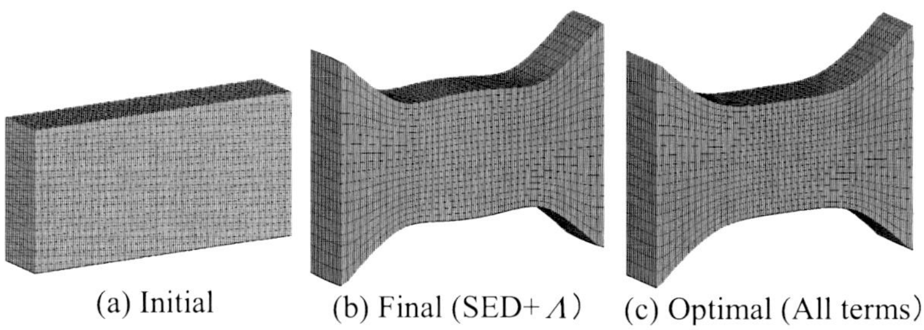

Figure 5: Calculated results of short span beam problem.

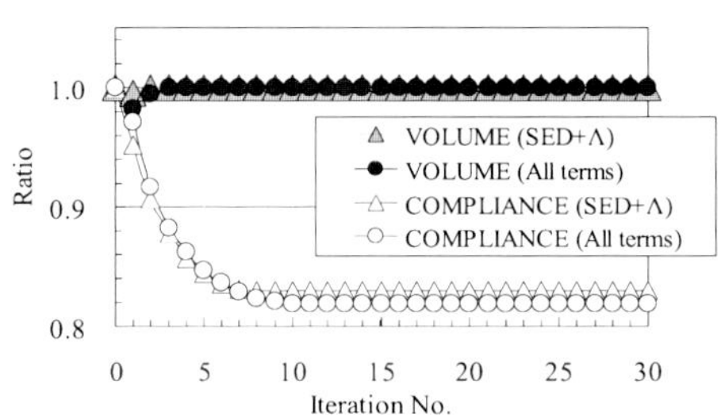

Figure 6: Iteration histories of short span beam problem.

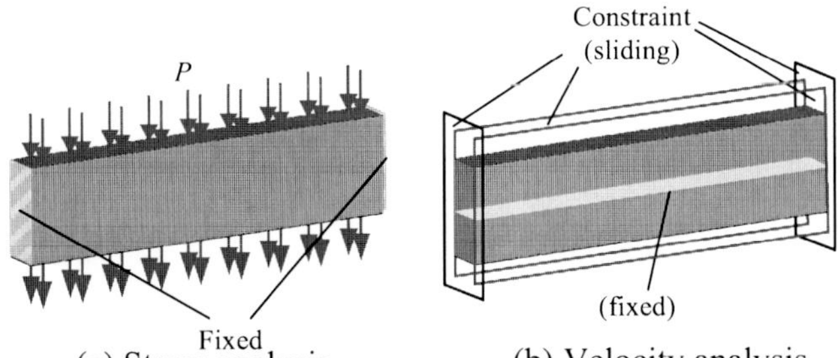

Figure 7: Boundary conditions for long span beam problem.

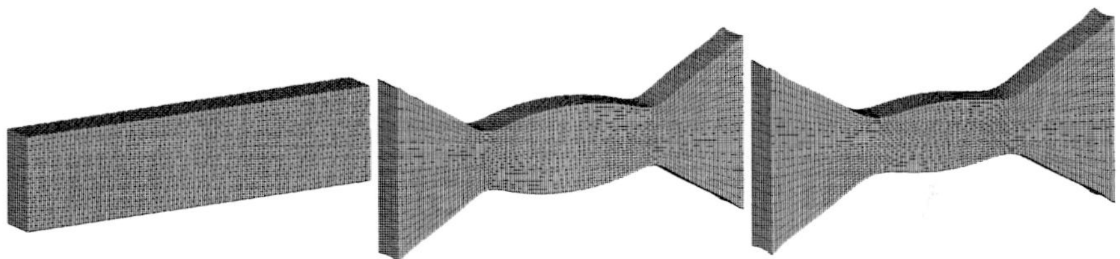

Figure 8: Calculated results of long span beam problem.

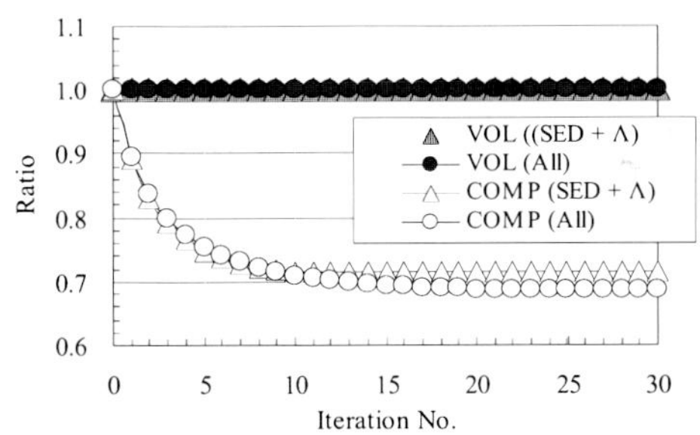

Figure 9: Iteration histories of long span beam problem.

5.2 Long span beam problem

A long span beam fixed at both ends was optimized. This long span beam is mainly subjected to bending stress. The boundary conditions and optimization conditions were the same as the short span beam problem as shown in Fig. 7(a) and (b). The initial shape and the optimal shape that was calculated by using eqn. (24) (except the term for body force $f_i(v_i + w_i)$) and eqn. (25) as the shape gradient function, are shown respectively in Fig. 8(a) and (c). For comparison, the final shape calculated by only using the strain energy density (SED) and Λ in eqn. (24) as the shape gradient function is shown in Fig. 8(b). It is seen that both shapes (b) and (c) obtained have smooth boundaries. It is also seen that the optimal shape (c) is clearly different from the shape (b). Figure 9 shows iteration histories of the volume and compliance in the optimization process. Both compliances are minimized while satisfying the volume constraint, but the rate of reduction is different; the rate for (b) is 28% while that for (c) is 32%. These results also confirm the effectiveness of the derived shape gradient function.

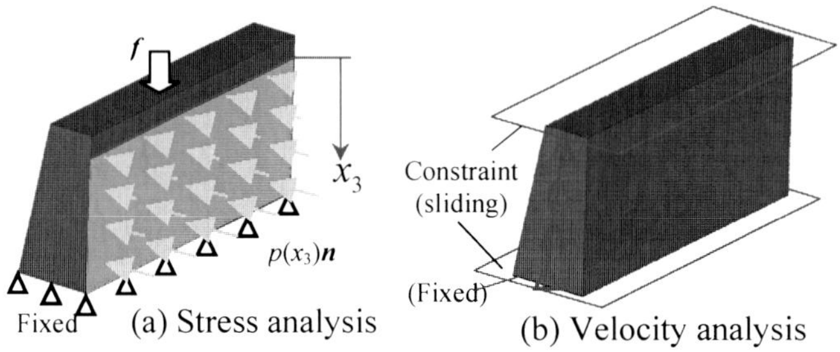

Figure 10: Boundary conditions of dam problem.

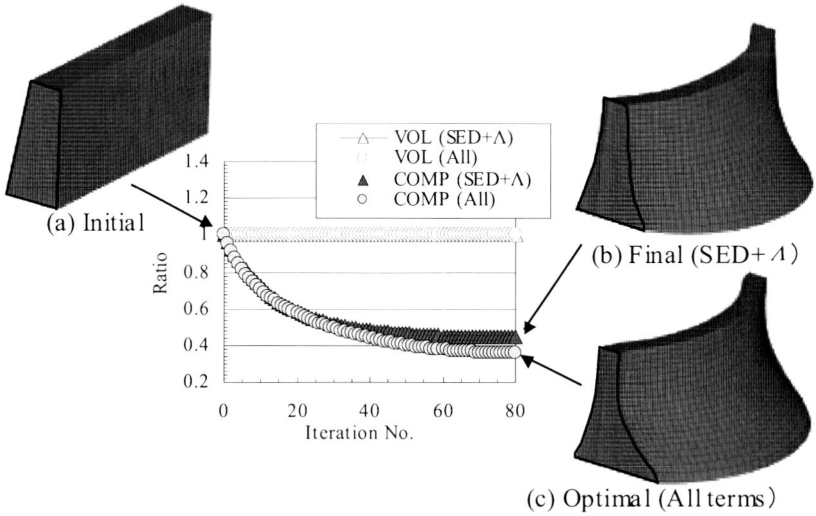

Figure 11: Iteration histories and obtained shapes of dam problem.

5.3 Dam problem

Boundary conditions for the shape optimization of a dam model are shown in Fig. 10. In the stress analysis, the bottom of the dam was fixed, and a hydro-pressure $p(x_3)\boldsymbol{n} = \rho g x_3 \boldsymbol{n}$ (ρ : density, g: gravity acceleration) and the gravity force $\boldsymbol{f}$ were applied (assuming $p'(x_3) = 0, \boldsymbol{f}' = 0$) as shown in Fig. 10(a). In the velocity analysis, the height and width were kept constant and the center lines of both side boundaries were fixed. The boundary with the hydro-pressure and the opposite boundary were set as the design boundaries to be optimised as shown in (b). A constant volume was set as the constraint. The iteration histories and obtained shapes are shown in Fig. 11, where (a) is the initial shape and (c) is the optimal shape obtained with eqns. (24) and (26) as the shape gradient function. To make the same comparison as in the beam problem, the final shape calculated by only using SED and Λ in eqn. (24) is shown in Fig. 11(b). It is seen that a smooth converged shape was obtained in each case and that the cross-section shapes are different. In both shapes, the objective functional decreased monotonically while satisfying the volume constraint. The compliance was reduced by 55% in (b) and by 64% in (c). These results also confirm the effectiveness our solution.

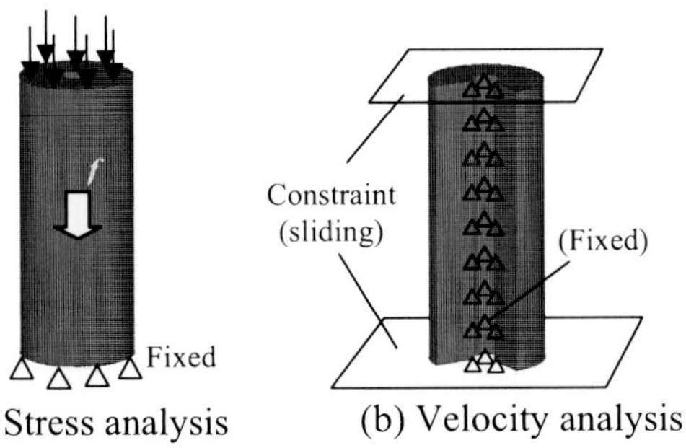

Figure 12: Boundary conditions of tower problem.

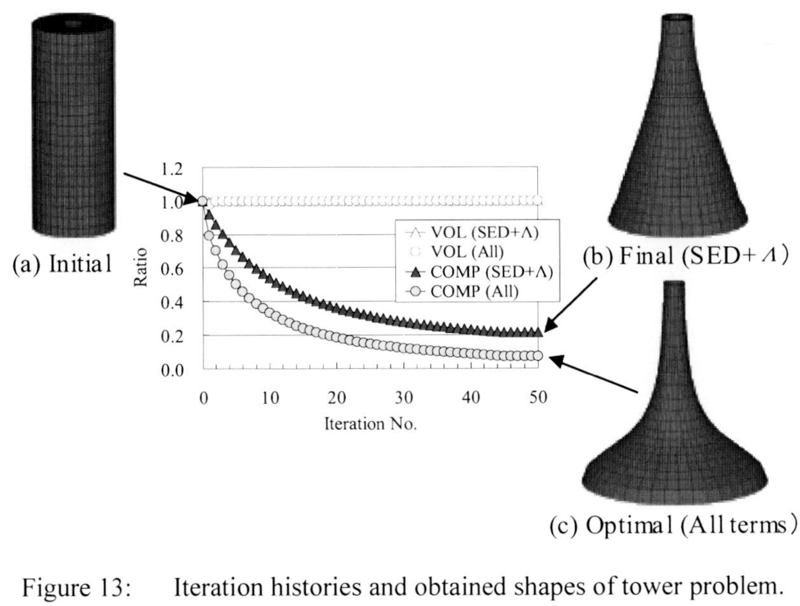

Figure 13: Iteration histories and obtained shapes of tower problem.

5.4 Tower problem

A cylindrical tower under the gravity force **f** and downward nodal forces at the top surface as shown in Fig. 12(a) was optimized. In the velocity analysis, the height was kept constant, and the inside hole was fixed as shown in Fig. 12(b). The volume remained constant as the constraint. The iteration histories and obtained shapes are shown in Fig. 13, where (c) is the optimal shape obtained by using all the terms in eqn. (24) as the shape gradient function. Depending on the magnitude of the shape gradient function, the upper portion contracted, and the lower portion was expanded. (b) is the final shape calculated by only using SED and Λ in eqn. (24) for comparison. The compliance was reduced by 80% in (b) and by 95% in (c) as expected, while satisfying the volume constraint.

6 Conclusion

In this paper we have proposed a practical solution to the shape optimization problem of solid structures. With this solution, the optimum free-form and smooth shape can be determined without any parameterization of the shape for optimization. For the rigidity design of a solid structure, a compliance minimization problem subject to a volume constraint is formulated as a non-parametric shape optimization problem. In order to reduce the total amount of applied loads, it is considered that the Neumann boundary is allowed to vary. The shape gradient function for this problem is derived, and is applied to the traction method. The validity and practical utility of this solution were verified based on the results obtained for four design examples.

References

[1] Azegami H. and Wu Z. C., Domain Optimization Analysis in Linear Elastic Problems (Approach Using Traction Method), *JSME Int. J.*, Series A, 39(2), pp. 272-278, 1996.

[2] Shimoda M., Azegami H. and Sakurai T., Traction Method Approach to Optimal Shape Design Problems, SAE Transactions, Journal of Passenger Cars, 106 (6), pp. 2355-2365, 1998.

[3] Shimoda M. and Tsuji J., Non-parametric Shape Optimization Method for Rigidity Design of Automotive Sheet Metal Structures, SAE Transactions, Journal of Passenger Cars, SAE Paper 200601-0584, 2006.

[4] Sokolowski J. and Zolesio J. P., *Introduction to Shape Optimization -Shape Sensitivity Analysis*, Springer-Verlag: New York, 1991.

[5] Haug E. J., Choi K. K. and Komkov V., *Design Sensitivity Analysis of Structural Systems*, Academic Press, Orlando, 1986.

[6] Azegami H., Kaizu S., Shimoda M. and Katamine E., Irregularity of Shape Optimization Problems and an Improvement Technique, in *Computer Aided Optimum Design of Structures V*, Hernandez V. S. and Brebbia C. A. (eds), Computational Mechanics Publications, Southampton, pp. 309-326, 1997.

Shape optimization of photoacoustic resonators

B. Kost, B. Baumann, M. Germer & M. Wolff
Department of Mechanical and Production Engineering,
Hamburg University of Applied Sciences, Germany

Abstract

The sensitivity of photoacoustic sensors strongly depends on the shape of the acoustical resonator. Up to now, mainly photoacoustic sensors consisting of a number of cylindrical parts have been investigated (cylinder cells, H cells, T cells etc.). In this paper, a numerical shape optimization of the resonator cell of photoacoustic sensors is described. The approach considers all shapes that can be represented by a number of axisymmetrical truncated cones which are connected in a continuous way. In addition, the geometry of the cell is subjected to certain constraints, e.g. the laser beam should not be blocked during its passage through the cell. The purpose is to maximize the sensor's signal strength. The acoustic pressure at the microphone of the sensor represents the objective function and is calculated using an eigenmode expansion combined with a finite element calculation. The solution of the 9-dimensional nonlinear optimization problem is a resonator shape with a substantial quality improvement with reference to the well-known H cell.
Keywords: photoacoustic, gas sensor, acoustic resonator, shape optimization.

1 Introduction

The photoacoustic effect, described by Bell as early as 1880 [1], can be used to develop sensors for trace gas analysis. There is a broad field of application for photoacoustic sensors, e.g. medical breath tests and environmental pollution control.

The general functional principle of a photoacoustic sensor is simple and is outlined in Figure 1. The sensor consists mainly of four parts: a photoacoustic cell, a microphone, a light source (laser) and a chopper. The cell contains a sample of the gas under investigation. The laser beam modulated by the chopper passes through a transparent window in the cell's front side. The laser wavelength is adjusted to an

WIT Transactions on The Built Environment, Vol 106, © 2009 WIT Press
www.witpress.com, ISSN 1743-3509 (on-line)
doi:10.2495/OP090051

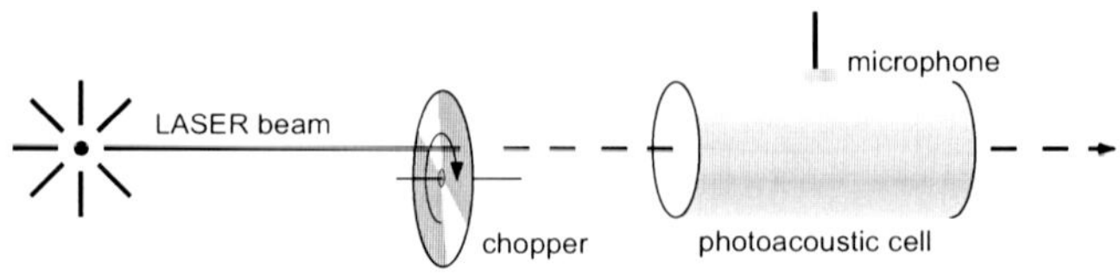

Figure 1: Design of photoacoustic sensor with cylindrical resonator cell.

absorption line of the trace gas causing molecular excitations accompanied by heat production. The periodical heat variation due to the chopper excites sound waves in the cell. The pressure field of the sound waves can be detected by the microphone and is equivalent to the signal strength of the sensor. The chopper frequency and the frequency of the wave propagation are identical. In order to amplify the microphone signal it is evident to adjust the chopper frequency to an acoustic resonance of the measuring cell, hence the name photoacoustic resonator. Details of the photoacoustic technique can be found in [2].

So far, mainly photoacoustic cells consisting of a number of cylindrical parts have been investigated. The most prominent example is a photoacoustic sensor with a cylindrical resonator cell. Another sensor design combines a cylindrical resonator with two cylindrical buffer volumes at the ends, known as H cell [3]. It is described in detail in section 3. Recently, the interest in photoacoustic sensors with unconventional design is increasing. The so-called T cell [4, 5] consists of two perpendicular arranged cylinders, one serving as the absorption cell, the other as resonator cell. A variation of the T cell is the X cell where the resonator cell has been split-up in two parts [6].

In order to further enhance the signal strength and sensitivity of a photoacoustic sensor in the following a numerical shape optimization of the resonator cell has been carried out. For that purpose the resonator cell is modeled by a number of cones. Details of the approach can be found in section 4.

2 Analysis of photoacoustic resonators

2.1 Calculation of the signal strength

Simple cylinder shaped resonators can be treated analytically. For cells of more complex shape, an analytical examination is impossible. Therefore, the investigation has to rely on numerical methods. Recently, a complete numerical description of the photoacoustic signal generation process including the computation of signal strength has been achieved [5, 7, 8]. In the following the numerical analysis of this approach is summarized.

In a first step the acoustical eigenmodes of the resonator cell have to be computed. The pressure field of the photoacoustical cell can be expressed by a weighted superposition of the acoustical eigenmodes. This leads to the amplitude resonance function of the resonator. According to experience, consideration of the first eight

Table 1: Gas parameters of N_2 under atmospheric conditions ($T = 293$ K, $p = 1013$ hPa) [11].

density ρ	$1.15\ \mathrm{kg/m^3}$
sound velocity c	$343\ \mathrm{m/s}$
viscosity η	$1.76 \cdot 10^{-5}\ \mathrm{Pa\,s}$
coefficient of heat conduction κ	$2.56 \cdot 10^{-2}\ \mathrm{W/m\,K}$
specific heat capacity at constant volume c_V	$0.741 \cdot 10^3\ \mathrm{J/kg\,K}$
specific heat capacity at constant pressure c_p	$1.038 \cdot 10^3\ \mathrm{J/kg\,K}$

eigenmodes has been found to be sufficient. The maximum value of the amplitude resonance function at a nontrivial resonance frequency can be considered to be the signal strength of the photoacoustical resonator cell.

For the evaluation of the amplitude resonance function the inclusion of loss is necessary. This has been done by means of a quality factor approach [9]. Stokes-Kirchhoff loss due to viscosity and thermal conduction as well as the thermal conductivity surface loss and the surface loss due to viscosity have been considered in this way.

The numerical calculations were carried out using the finite element tool COMSOL Multiphysics 3.4 [10]. The results were obtained with the default solver parameters.

2.2 Basic data

In the following simulations it is assumed that the cells are filled with nitrogen N_2 with traces of NO under atmospheric conditions. The gas parameters of N_2 are taken from [11] and can be found in Table 1.

To obtain proper laser data an experimental investigation has been carried out. The investigation of our set-up reveals that the cross-sections of the laser beam intensity are nearly circularly Gaussian profiles. The effective radius has been found to be approximately 1.7 mm independent of the axial position.

The intensity decrease in the longitudinal direction is given by an exponential function using a decay factor α. The absorption coefficient of NO which is responsible for the decrease of the laser intensity along the cell axis is $\alpha = 0.166\ \mathrm{m}^{-1}$ [12]. This quantity represents also a measure of the NO concentration in the cell.

3 Reference cell

For the numerical evaluation of the solution quality of the shape optimization described below the well-known H cell [3] has been taken as a reference. The small-sized cylinder of diameter d_1 in the middle represents the resonator. At both ends are larger cylinders of diameter d_2 serving as buffer volumes. A sectional

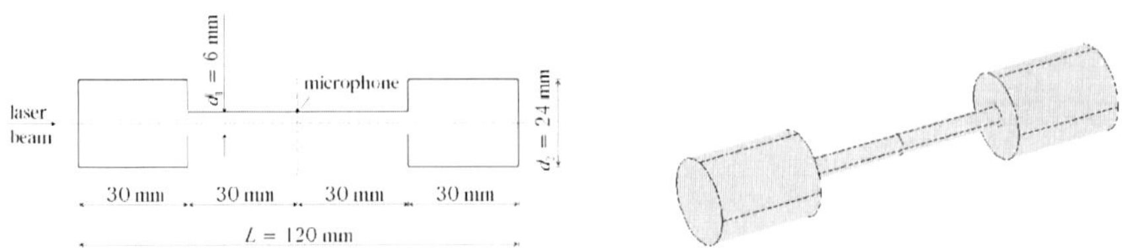

Figure 2: Reference cell: design data (left), 3D model (right).

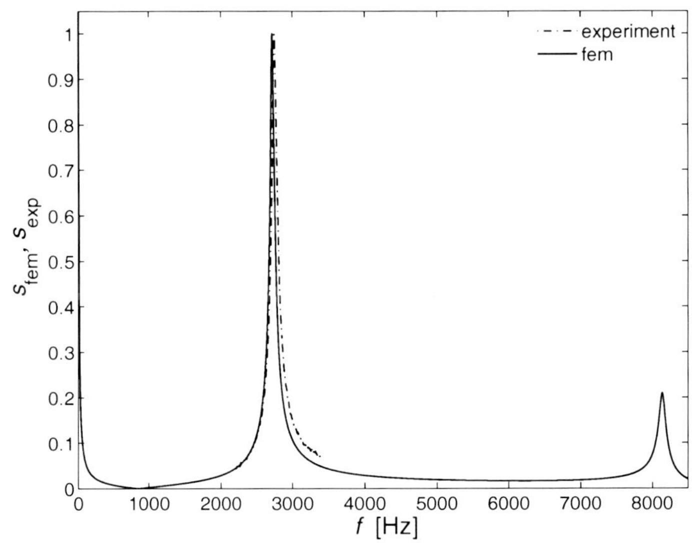

Figure 3: Comparison of normalized amplitude resonance function $s_{\text{fem}}(f)$ and microphone signal function $s_{\text{exp}}(f)$ for the reference cell over the numerically analyzed frequency range.

drawing of the reference cell with the design data used in this investigation is shown in Figure 2 (left). A 3D model can be found in Figure 2 (right).

The quality of the reference cell has been measured experimentally in our laboratory over a certain range of the chopper frequency f. Additionally, the cell has been analyzed by the finite element eigenmode expansion approach described in section 2 over a larger range of f. This has been done to cover a sufficient number of eigenmodes in the series expansion. Both results, the experimentally obtained microphone signal function $s_{\text{exp}}(f)$ and the numerically calculated amplitude resonance function $s_{\text{fem}}(f)$, are depicted in Figure 3. To allow a reasonable comparison the data have been normalized with respect to the maximum value of each function.

The numerical and experimental data are in good accordance. The frequency of the experimental signal maximum is 2753 Hz, the frequency of the numerical

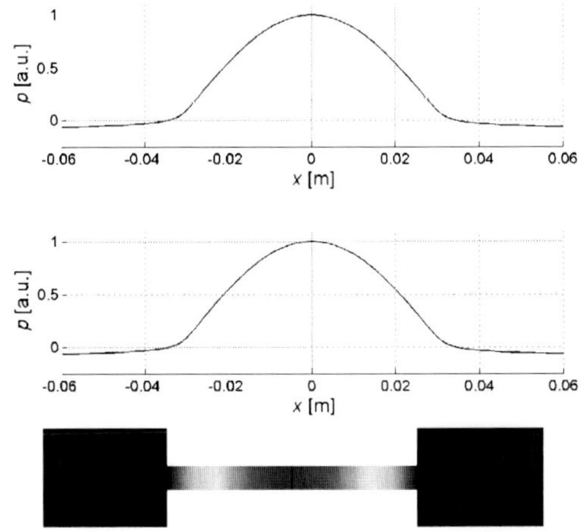

Figure 4: Dominant eigenmode of the reference cell (2nd nontrivial mode at 2721 Hz), acoustic pressure distribution p along cell axis x (top), mode shape $|p|$ (bottom).

resonance maximum is 2721 Hz. This is a relative deviation of approximately 1.2 %. The graphs of the functions are quite similar. The width at the peak of the experimental function is slightly larger than that of the numerical function. Since the width of the function is determined by loss effects and the finite element simulation does not include all effects that contribute to loss of the photoacoustic signal, this is an expected characteristic. However, the good accordance of the numerical and experimental data shows that the most important loss effects have been considered in the finite element model. Further investigations concerning the comparison of analytical, experimental and finite element results of certain photoacoustic cells can be found in [5, 7].

In Figure 4 (bottom) the dominant eigenmode of the amplitude resonance function is shown, the 2nd nontrivial mode. It consists mainly of a longitudinal mode of the center cylinder. The according eigenfrequency of a cylinder with both ends open and a length of 60 mm can be analytically determined to 2858 Hz. Figure 4 (top) shows the acoustic pressure distribution along the cell axis. The distance > 60 mm between the zero crossings of the function is equivalent to an effective length of the middle cylinder, hence the lower frequency 2721 Hz. More about end corrections can be found in [13].

4 Shape optimization model

The finite element model for shape optimization of a photoacoustic cell consists of 8 axisymmetrical truncated cones of equal length $\frac{L}{8}$. The cones are connected in a continuous way. The 9 radii $x_i (i = 1, 2, \ldots, 9)$ of the cones represent the optimization variables. In Figure 5 a 3D model of a potential resonator cell is shown.

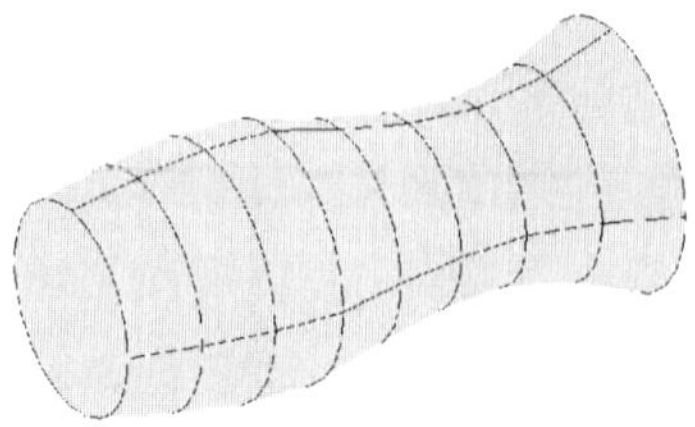

Figure 5: 3D model of a potential resonator cell contained in the search space.

The objective function

$$z : \mathbb{R}^9 \longrightarrow \mathbb{R}\,, \quad \boldsymbol{x} \longrightarrow z(\boldsymbol{x}) =: Z \tag{1}$$

of the shape optimization problem is set to $z := |p_\mathrm{m}|$. The function p_m determines the value of the pressure field at the microphone position. This is a measure of the signal strength of the photoacoustic cell. According to this, the objective function value Z can be calculated by the procedure described in section 2.1.

The optimization parameters have to fulfill the inequaltities

$$r_\mathrm{lb} \leq x_i\,, \qquad i = 1, 2, \ldots, 9 \tag{2}$$

where $r_\mathrm{lb} = 3$ mm is a lower bound for the radii. The value r_lb is sufficiently larger than the laser beam radius. These constraints have been imposed to prevent that the laser beam is blocked during its passage through the cell and to ensure a proper energy input into the system. Due to technical reasons the radii have also been restricted by upper bounds

$$x_i \leq r_\mathrm{ub}\,, \qquad i = 1, 2, \ldots, 9 \tag{3}$$

with $r_\mathrm{ub} = 60$ mm. This guarantees a compact design of the photoacoustic cell.

With the upper and lower bounds the subset $\mathbf{X} \subset \mathbb{R}^9$ defined by

$$\mathbf{X} := \left\{\boldsymbol{x} \in \mathbb{R}^9 \,|\, r_\mathrm{lb} \leq x_i \leq r_\mathrm{ub}\,, i = 1, 2, \ldots, 9\right\} \tag{4}$$

is the feasible region. A solution of the nonlinear maximization problem can be expressed by

$$\boldsymbol{x}^* := \operatorname{argmax}\left\{z(\boldsymbol{x}) \,|\, \boldsymbol{x} \in \mathbf{X}\right\} \quad \text{and} \quad Z^* := z(\boldsymbol{x}^*)\,. \tag{5}$$

5 Optimization results

The shape optimization problem of section 4 has been solved with the MATLAB [14] standard optimization function for constrained problems `fmincon`. Preliminary investigations show that the shape optimization problem of photoacoustic resonator cells may have multiple local optima [15]. Therefore, the optimization has been rerun with different initial shapes, representing different starting points in the search space. For all initial shapes the smallest radii have been set to 3 mm, the largest to 12 mm.

It seems obvious to start the optimization at a cylindrical resonator shape. However, this is problematic because due to the axial symmetry combined with an approximate translation symmetry of the laser intensity distribution in a cylindrical resonator only radial modes can be excited. The low value of the absorption coefficient and the small length of the cell causes only a marginal decrease of the laser intensity of about 2 %. Therefore, an effective excitation of longitudinal or azimuthal modes is not possible. In practice impurities in the gas, deviations from the symmetry due to the laser adjustment and other factors lead to a weak excitation of nonradial modes. However, the unsatisfying performance of the cylinder cell can be explained by the facts mentioned above. The eigenfrequencies of the radial modes are very high and with respect to the numerical investigations of the photoacoustic sensor beyond the interesting frequency range.

A collection of initial shapes used in this investigation and resulting solutions are depicted in Table 2 together with the ratio

$$\zeta := \frac{Z^*}{Z_{\text{ref}}} . \tag{6}$$

The parameter ζ specifies the solution quality in comparison to the reference cell. Z_{ref} is the maximum value of the not normalized amplitude resonance function of the H cell resonator.

Even though the decrease of laser intensity along the cell axis is weak, it could not be excluded that it effects the appearance and quality of the solution. Therefore, albeit similar, both geometries (cases 1 and 2 of the table) have been chosen as initial shapes. However, the obtained solutions are similar too and of poor quality. Case 3 has been included in the collection of initial shapes with the intention to have a starting point which is in the limits of our modeling close to the reference cell. The result shows a decent improvement of the signal strength of about 20 %. Case 4 and 5 have been chosen to provoke the exploration of different parts of the search space. Although, the solution geometry of case 4 looks rather different compared to case 3 the ζ-value is only slightly better. The best solution quality exhibits case 5. Compared to the reference cell an improvement of about 31.6 % could be found. In Figure 6 a 3D model of the solution is depicted. The best geometry found is an asymmetrical shape with horn-like ends.

In Figure 7 (bottom) the dominant eigenmode of the amplitude resonance function of the best shape is shown, the 4th nontrivial mode. Figure 7 (top) shows the acoustic pressure distribution along the cell axis.

6 Conclusions

In the present article we described the numerical shape optimization of the resonator cell of axisymmetrical photoacoustic sensors. The shape optimization problem of such sensor cells is nonconvex, hence multiple local optima exist. To ensure

Table 2: Longitudinal sections of initial shapes and solutions. ζ is the solution quality ratio, see Eq. (6).

No.	initial shape	solution	ζ
1	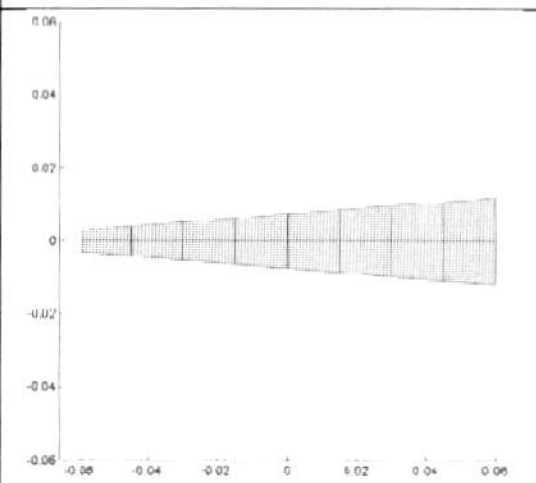	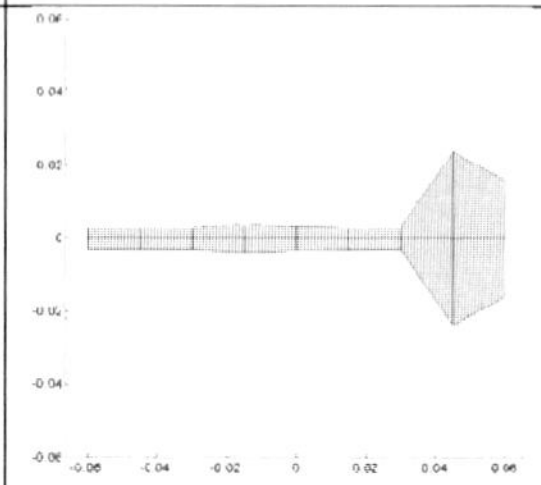	0.890
2	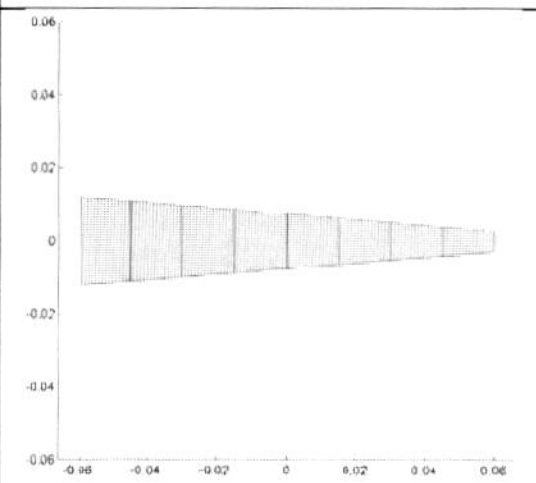	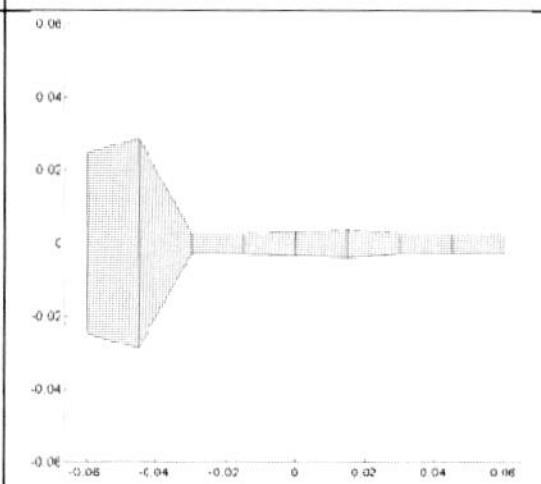	0.941
3	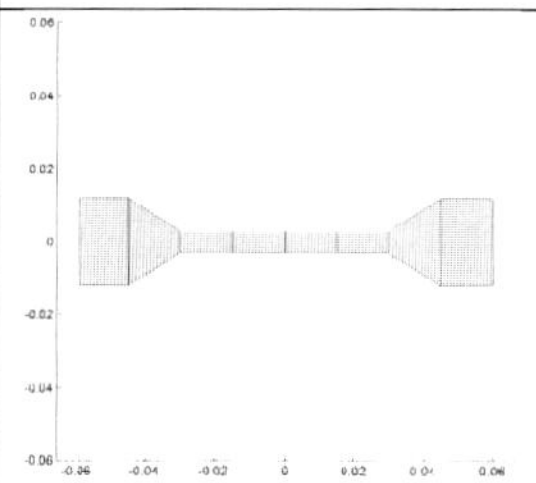	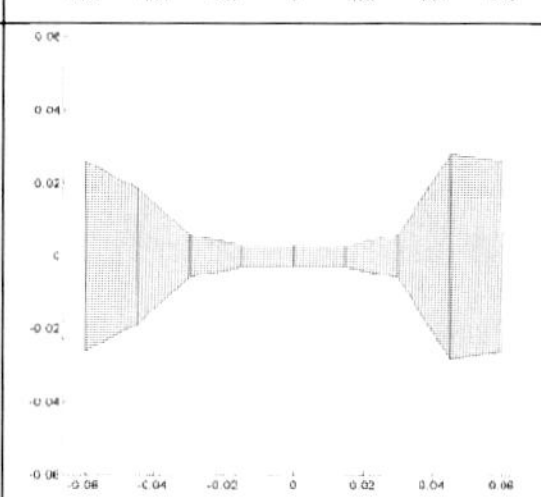	1.198
4	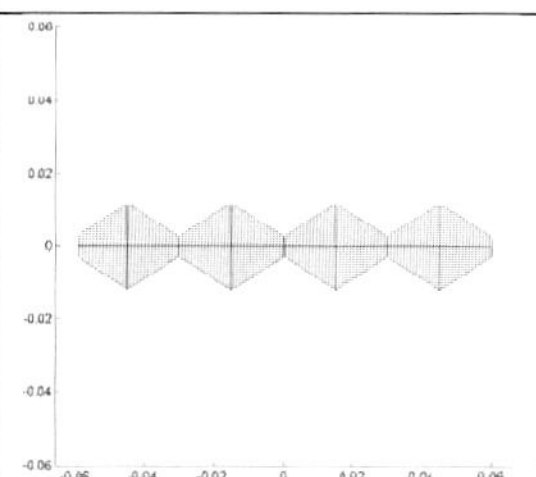	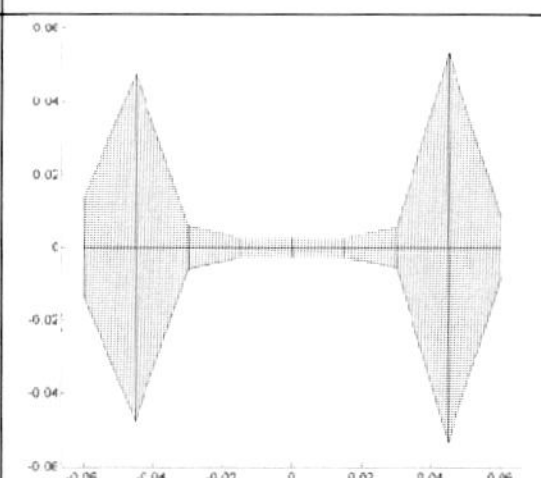	1.219
5	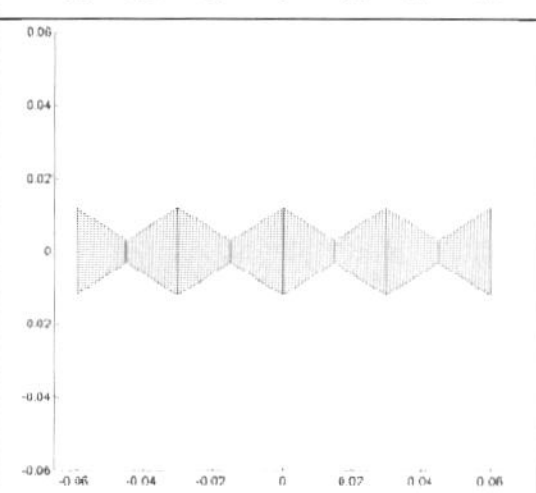	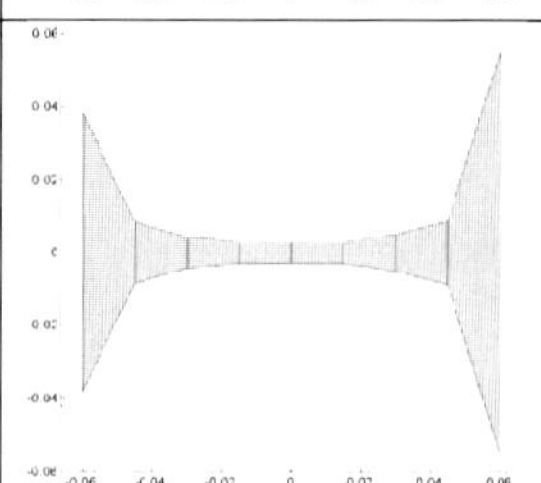	1.316

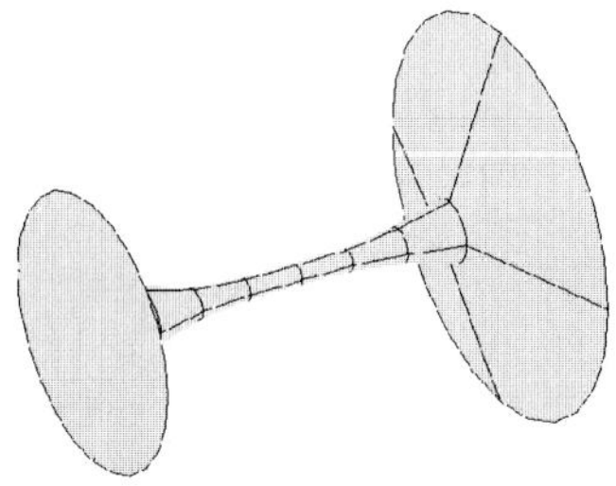

Figure 6: 3D model of best resonator shape.

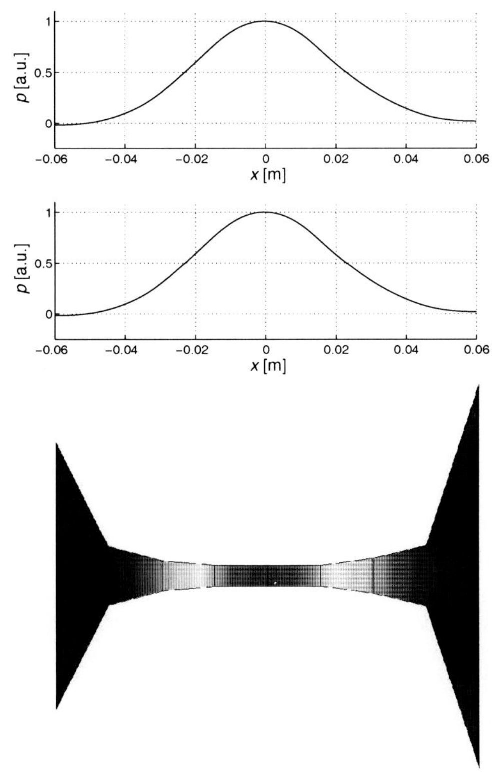

Figure 7: Dominant eigenmode of best cell shape (4th nontrivial mode at 2673 Hz), acoustic pressure distribution p along cell axis x (top), mode shape $|p|$ (bottom).

an adequate exploration of the search space a collection of initial shapes have been used. The best solution found is a resonator shape with horn-like ends. With respect to the well-known H cell this cell shows a considerably improvement in signal strength of about 32 %. An experimental verification of this result is planned.

References

[1] Bell, A.G., On the production and reproduction of sound by light. *Am. J. Science*, **20**, pp. 305–309, 1880.

[2] Michaelian, K.H., *Photoacoustic Infrared Spectroscopy*. Wiley-Interscience: Hoboken, 2003.

[3] Nodov, E., Optimization of resonant cell design for optoacoustic gas spectroscopy (H-type). *Applied Optics*, **17**, pp. 1110–1119, 1978.

[4] Wolff, M., Groninga, H.G., Baumann, B., Kost, B. & Harde, H., Resonance investigations using PAS and FEM. *Acta Acustica*, **91**, pp. 1477–1481, Suppl. 1, 99, 2005.

[5] Baumann, B., Kost, B., Groninga, H. & Wolff, M., Eigenmode analysis of photoacoustic sensors via finite element method. *Review of Scientific Instruments*, **77**, 044901, 2006.

[6] Kost, B., Baumann, B., Wolff, M. & Groninga, H., Numerical investigations of photoacoustic sensor signals. *Acta Acustica united with Acustica*, accepted for publication, 2009.

[7] Baumann, B., Wolff, M., Kost, B. & Groninga, H., Finite element calculation of photoacoustic signals. *Applied Optics*, **46**, pp. 1120–1125, 2007.

[8] Baumann, B., Wolff, M., Kost, B. & Groninga, H., Solving a coupled field problem by eigenmode expansion and finite element method. *International Journal of Multiphysics*, **1(3)**, pp. 303–315, 2007.

[9] Kreuzer, L.B., The physics of signal generation and detection. *Optoacoustic Spectroscopy and Detection*, ed. Y.H. Pao, Academic: London, pp. 1–25, 1977.

[10] COMSOL-Homepage: www.comsol.com.

[11] *VDI-Wärmeatlas*. 9. Auflage, Springer-Verlag, 2002.

[12] Rothman, L.S. et al., The HITRAN 2004 molecular spectroscopic database. *Journal of Quantitative Spectroscopy & Radiative Transfer*, **96**, pp. 139–204, 2005.

[13] Miklós, A., Hess, P. & Bozóki, Z., Application of acoustic resonators in photoacoustic trace gas analysis and metrology. *Review of Scientific Instruments*, **72(4)**, pp. 1937–1955, 2001.

[14] MATLAB-Homepage: www.mathworks.com.

[15] Baumann, B., Kost, B., Wolff, M., Groninga, H., Blöß, T. & Knickrehm, S., Numerical shape optimization of photoacoustic sample cells: First results. *Proceedings of the European COMSOL Conference 2007*, Grenoble, France, pp. 1–6, 2007.

An innovative approach to the design of multi-bolted joint assemblies in the uncertain early stages

T. Dericquebourg[1], D. Lenoir[2], F. Debarle[1] & L. Jézéquel[2]
[1]*PSA Peugeot Citroën, Centre Technique de La Garenne-Colombes, France*
[2]*Laboratoire de Tribologie et Dynamique des Systèmes, Ecole Centrale de Lyon/Université de Lyon, France*

Abstract

In the automotive industry, the optimization of the preliminary stages is a major asset for reducing the time of the design process. So, a new approach for setting up the optimal configuration of a bolted joints assembly has been developed. The aim is to efficiently fix the main characteristics of bolted joint assemblies in a particularly uncertain context. Indeed, in preliminary design, choices are made with little information and many indeterminations, but these should be relevant whatever the following steps of the design process may be. The methodology developed is based on simplified models included in a strategy consisting of testing configurations to choose the best one. In order to scan the design space well, an optimized Latin hypercube strategy is used. Then, algorithms have been developed to find clusters of good items and of unacceptable ones. These clusters are then used to specify a robust solution or to avoid incorrect configurations. An application was conducted on an automotive mounting. The reduced time needed to obtain useful results really contributes to shortening the design process duration and the approach allows one to control the risks of the process uncertainties.

Keywords: mechanical design, preliminary design, bolted joints, uncertain context, modelling.

WIT Transactions on The Built Environment, Vol 106, © 2009 WIT Press
www.witpress.com, ISSN 1743-3509 (on-line)
doi:10.2495/OP090061

1 Introduction

Time is one of the main industrial constraints for manufacturers, who are continually seeking to improve the efficiency of the first stages of the design process. The decisions made in the preliminary stages considerably affect the following ones and, consequently, the due time criteria. This aspect is predominant in the automotive industry since time reduction is an asset for efficient performance in the current competitive market. This aspect is reflected in all the fields of the car manufacturing hierarchy, in the acoustical domain for example, where engineers try to quickly assess the acoustical performances of design alternatives using meta-models to improve space allocations [1]. It is also particularly true in bolted adjoint assemblies. Such assemblies are widespread in the automotive industries since they are very practical for setting up removable junctions between the subassemblies of the body structures or unit designs. Designers always seek to specify the number, the position or the main characteristics of bolted joints quickly and efficiently. In this framework, it may be too risky to rely only on designer experience and on simple empirical formulas based on simplified hypotheses. Too basic an approach may well be of little help since it may not take into consideration the most significant parameters and complex constraints for some problems.

The object of this paper is to present a new approach to assist designers in making the most appropriate choices in the first stages of the bolted joints definition process. This paper outlines the design of screwed assemblies, and thus the optimization of the main parameters at the higher level of the design cycle to satisfy the maximum mechanical criteria. The study was conducted on industrial cases and the methodology has been developed particularly on mountings made up of subassemblies linking the engine to the body shell with screwed joints. An example of mounting can be seen in figure 1 – this is a common bolted assembly in car industries.

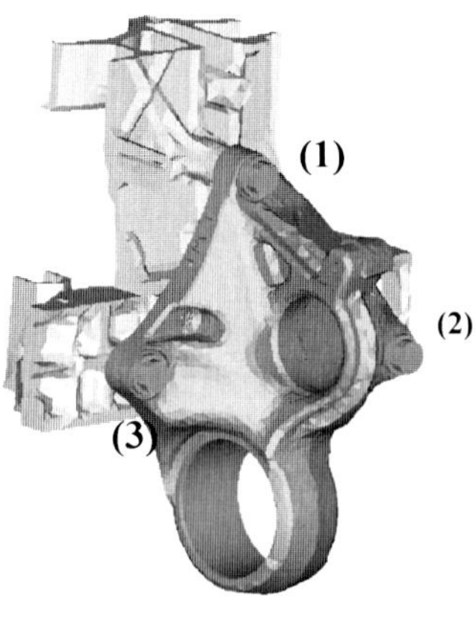

Figure 1: Example of a mounting fastened to an engine crankcase with three bolts.

2 Specific context

The specific context of this study has a significant impact on the way to settle the problem. Indeed, the methodology has been developed in order to be used in the early stages of the design process. In these early stages, the context is very uncertain because almost no information is known, or if information is given, it is not clear or is prone to alteration. Moreover, the decisions made in the first steps of the design process can have a real impact on the following design steps and so on the design cycle time.

In preliminary design, the aim of multi-bolted joint assembly pre-dimensioning is to determine a good configuration that will fulfil several requirements. A configuration of a bolted joints assembly is defined by the main characteristics of the assembly and the main requirements are essentially the screw strength, preventing mating of parts in contact, preventing sliding and preventing detachment. However, in this context, the uncertainties are multiple and of several kinds. The aim of the methodology is to suggest some solutions that are able to fulfil the needs and give some accurate information about the capabilities of the solutions to satisfy the criteria in order to help the designers to efficiently make the good decisions as early as possible in the process.

2.1 Uncertain input data

In preliminary design, input data are not well defined. Indeed, at the first stages of the design process, the different data are estimated but not accurately calculated: for example, the loads or some geometrical characteristics of the surroundings parts cannot be precisely known at this step, but can only be estimated thanks to previous conceptions. As all the external characteristics of a part, such as loads or common contacts, cannot be accurately identified, it is difficult to correctly design the assembly and to be sure that the created design will fit the needs. Moreover, while the part is not completely defined, it is not easy to determine if this part will fulfil the requirements to declare the part well designed or not. As a result, the main difficulty in this context is to be able to recommend a good solution that fits the needs when not all surrounding parameters are well known. This makes one of the main difficulties of the problem.

2.2 Following design step uncertainties

The first choices made during the beginning of the design process are very important because they can have high consequences on the following steps. Indeed, once the choice of a solution is made, all the following steps are consequences of this first choice. As a result, this first choice must absolutely be a good one in order to meet the deadlines of the design process and to avoid many loops between the design steps that are time consuming. In this context, for an adequate design, choices made at the first stages should be relevant during the following stages. This criterion is justified by the fact that the subassemblies may be designed and often developed in different departments and, therefore, by different engineers in different design offices. Thus the main parameters of an assembly, which have a real impact on the design of the subassemblies, have to be established as soon in the process as possible. The main difficulty in this context is to be sure that the solution will satisfy all the requirements as the design process goes along, because the consecutive steps are partly unpredictable.

3 Methodology

The methodology is based on simplified models that have been specifically developed. To develop a simplified model, several hypotheses have to be done and consequently restrict the validity field. In this case, the hypotheses retained to develop the simplified model are the following: the clamped part must be composed of a stiff centre area and flexible edges where the bolts are placed. Using the existing models found in the literature, such an assembly cannot be fully represented and thus cannot be efficiently designed. The best approach remains FE modelling. Unfortunately, this is heavy and time consuming, and thus inappropriate in an early phase of the design process. This is why the presented simplified modelling has been developed. Then, the methodology consists of using the simplified model in an iterative approach as a tool for choosing the optimal configuration for a bolted industrial assembly.

3.1 General approach

The principal objective of this methodology is to set up the main parameters and characteristics of bolted joint assemblies subjected to static or cyclic loads in normal and extreme operation. The mains parameters are the following:

- number of screws,
- position of bolted joints in the assembly,
- characteristics of the bolts (screw diameter, length, pre-load, etc…).

However, the main industrial constraints in the first stages of a design process are rapidity and simplicity. Hence, the methodology developed is based on simplified, easily executable models in the Matlab® environment. The key idea of the resolution procedure is to iterate on several potential configurations so as to find the one that best fits the various design criteria.

In this methodology, an architecture is defined by the positions and the number of screws. When geometrical characteristics (screw and local clamped part dimensions) and material properties are added, this defines a configuration (cf. figure 2). In the aim of investigating all the possible solutions, the designer must specify the following different input parameters:

- load values,
- location of the external loading,
- vacant geometric areas for screw implementation,
- vacant space for material filling.

Below, all possible configurations are investigated to set up stresses in the joints and bolts for each case. Finally, the results are compared and the optimal configuration is chosen by the designer. The described process highlights the necessity for very light, quick models.

3.2 Simplified model

The simplified model used in the methodology has been previously developed and presented in [2]. The simplified model is made up of two levels: a local level dedicated to describe the local behaviour of a bolted link and a global level dedicated to compute the forces repartition between the bolts.

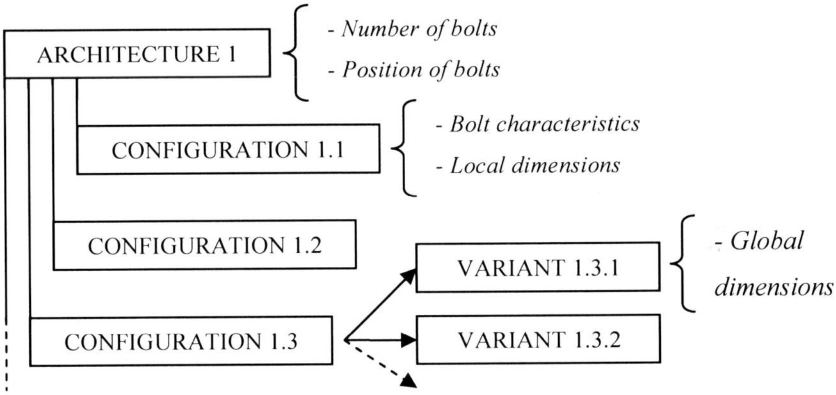

Figure 2: Solutions classification: a variant is completely defined and is based on the configuration of an architecture.

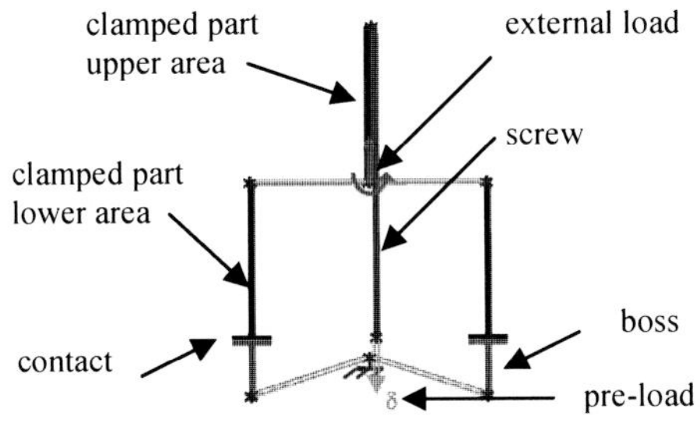

Figure 3: Simplified local model representation (longitudinal-section).

3.2.1 Local level

Concerning the local approach, a new model has been developed to overcome the limits of the existing models. The model can take complete or partial separation of the subassemblies into consideration under normal and extreme operating conditions. The local model represents a bolted joint consisting of a screw, its corresponding boss and the clamped subassembly (cf. Figure 3). It represents only the local zone of the parts, located near the screw. The inputs are the external load set up from the global model and the joint and screw geometric and physical characteristics. This model is essentially made up from simple finite elements like beam or rigid elements and node to node contact. The stiffness of the constitutive elements is fixed to physical values partially calculated thanks to the VDI [3] that are the most familiar references used in industrial design. But other adaptations have been made in order to be more predictive.

3.2.2 Global level

The main aim of the global level is to model the assembly using simple finite elements like rigid elements and beam elements; this model allows calculating the forces repartition between the joints that is function of the clamped part shape. This model can be described as a skeleton, since the different bolted joint centres are linked to the load application point by rigid and beam elements: rigid elements are used to model the stiff areas of a subassembly whereas beam elements are used to describe the flexible zones (cf. Figure 4). The characteristics of beam elements are established to fit the shape and stiffness of the subassemblies. Material properties are the same as in the real subassembly.

3.3 Solutions searching strategy

Starting with the input data listed previously, a listing of different architectures can be constructed. The strategy consists then in trying these several architectures and for each, testing several configurations. The robustness of each configuration is then estimated thanks to the simplified model. All parameters, except the ones concerning the global shape of the clamped part, are fixed in a configuration. When these last parameters are fixed at a defined value, the entire simplified model is defined and called a variant (cf. Figure 2).

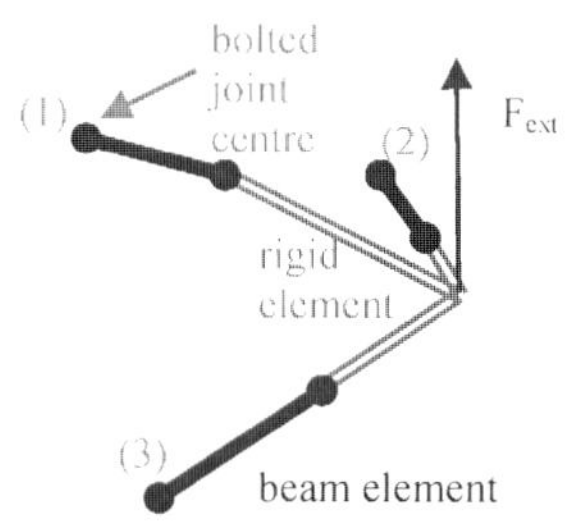

Figure 4: Equivalent global modelling of a 3-bolts assembly.

Table 1: Failure risks for every mechanical criterion and global failure risk for each configuration of the same architecture (illustration).

Config.	Criterion 1	Criterion 2	...	Criterion n	Global criterion
1.1	10%	40%		30%	52%
1.2	12%	16%		25%	45%
1.3	8%	23%		14%	25%
...	...				

To estimate the viability of a configuration, some variants of the configuration are generated (at least a hundred) using an optimised Latin hypercube [4–6] tactic in order to well scan the design space. Each variant is calculated in order to check if the variant fulfil the mechanical criteria. Then a kind of failure risk can be given when calculating the ratio between the number of faulty variants and the number of tested variants. This calculation is done for each criterion of acceptance allowing determining which criterion is the most difficult to satisfy. Then, a table can be created to quickly show the results and the probabilities for each configuration of all possible architectures (cf. Table 1).

The simplified model allows testing all criteria that have to be checked in order to determine if a solution is acceptable or not. The main criteria tested are the followings:

- mechanical axial resistance,
- bending resistance,
- shear resistance,
- slip resistance,
- no detachment,
- fatigue resistance.

And finally, a global acceptance criterion is calculated, it represents the global failure risk for all the criteria together. This last criterion is not satisfied if just one of the other criteria is not satisfied.

This table is very useful to sort the different configurations in failure risk order: the lower the failure risk is, the easier the design process will be, but in the same time, the more expensive the assembly conception may be. On the other hand, if the failure risk is high, there will certainly be difficulties in the design process to obtain a viable assembly. The choice of the configuration must so be a compromise between the performance of the assembly (i.e. the capability of the assembly to satisfy the various mechanical criteria) and its cost (the design process cost as well as the production cost).

3.4 Solutions gathering algorithm

For each solution, useful information is given: the failure risk for each criterion. So if this risk is very weak, it is not too hazardous to choose this solution, but it will certainly be more expensive. It is more interesting to try to find a less expensive solution if it is possible to well specify it. Indeed, a solution can have a non neglecting failure risk but if you are able to well describe the parameters associations that lead to this failure or the ones that allow avoiding failure, it is much more interesting, because it is far possible that this solution will be cheaper. As said previously, the aim is to find the good compromise between the cost and the performance of the solution. And if specifications can be done to avoid failure areas, it will be possible to choose more risky solutions and obtain a cheaper solution with a risk control.

In order to find the design space areas to avoid or to use allowing specifications, two similar algorithms have been developed. One dedicated to find the clusters of good solutions and the other to find the clusters of bad solutions.

The diagram in figure 6 exposes the algorithm structure and an example illustration can be shown on figure 5. The initialisation of the algorithm consists in finding the first valid solution (1) that is the farthest of the bad ones:

$$X_1^{cluster} = \max_{i \in I}\left(\min_{j \in J}\left(X_i^{ok} - X_j^{ko}\right)\right) \tag{1}$$

where $X_1^{cluster}$ barycentre of the cluster number 1

X_i^{ok} variant number i of the valid variants set

X_j^{ko} variant number j of the unacceptable variants set

I set of valid variant indexes and J set of unacceptable ones

Then, the iterative procedure of the algorithm works in that way: from the first good solution (1), the algorithm seeks for the nearest good solution (2), and with these two solutions a new cluster is created; a cluster is defined by the included solutions and the barycentre $X_1^{cluster}$.

$$X_1^{add} = \min_{i \in I}\left(X_i^{ok} - X_1^{cluster}\right) \tag{2}$$

where X_1^{add} next variant to include in the cluster 1

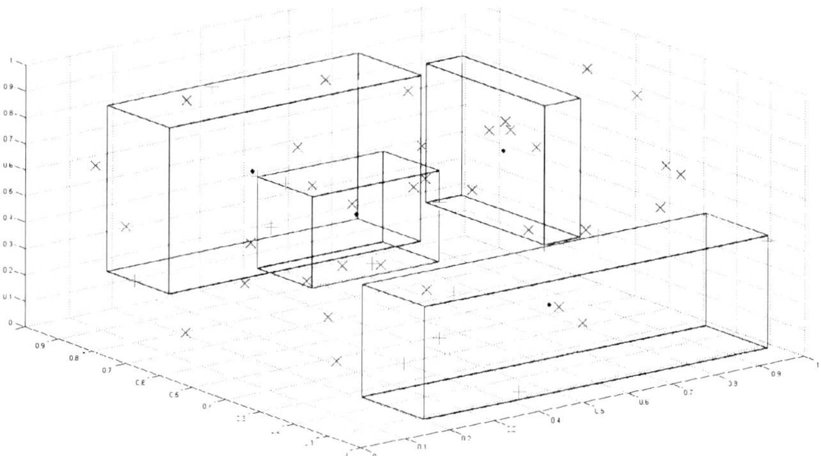

Figure 5: Example in 3-dimensions of good solution clusters extracted with the algorithm (+ : good solution, x : bad solution).

Then a test is made: if no bad solution is included inside the limits of the cluster, the cluster generation goes on by seeking again the nearest good solution (2) and so on…, else the cluster growth ends. So the nearest good solutions are iteratively included in the cluster until the cluster is going to encircle a bad solution. In this last case, the cluster growth ends and a new one is created by beginning with another good solution (3) determined like the first one (the farthest good solution from the bad ones) but here excluding the already grouped solutions. Then the growth of the cluster (4) is similarly performed.

$$X_k^{cluster} = \max_{\substack{i \in I \\ i \notin Cp, p<k}} \left(\min_{j \in J} \left(X_i^{ok} - X_j^{ko} \right) \right) \tag{3}$$

$$X_k^{add} = \min_{\substack{i \in I \\ i \notin Cp, p \leq k}} \left(X_i^{ok} - X_k^{cluster} \right) \tag{4}$$

where $X_k^{cluster}$ barycentre of cluster number k

X_k^{add} next variant to include in the cluster p

Cp set of variant indexes included in the cluster p

k current cluster number

The second algorithm dedicated to group the defective variants has a similar working. Indeed, the structure is the same, except that the valid solutions are replaced by the bad ones and inversely.

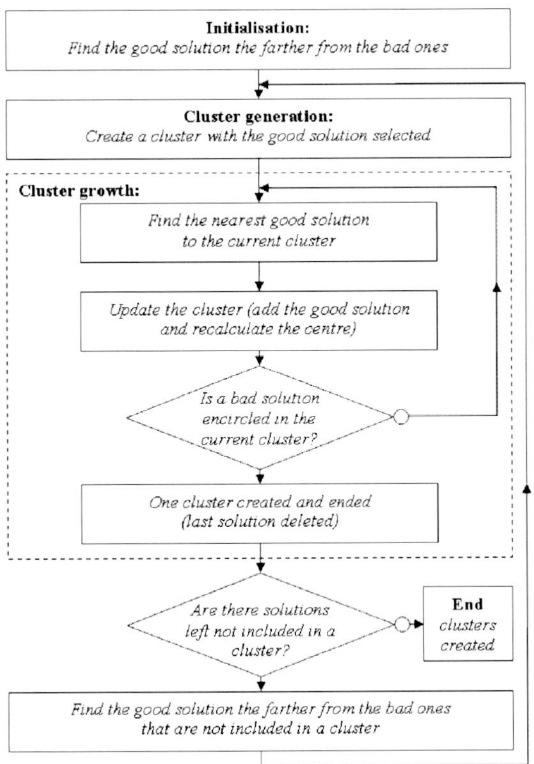

Figure 6: Algorithm structure.

4 Industrial application

4.1 Application case

The methodology presented was directly applied to different mountings that are subassemblies linking the engine to the body shell of a car with bolted joints as shown on figure 1. The main function of these parts, fixed to the engine with several bolts, is to hold the latter in the car and to prevent its excessive movement. Generally, the mountings have similar shapes that satisfy the hypotheses of the simplified modelling: the core is often dense and rigid, while the edges are more flexible.

For the presented industrial application, two architectures, a 3 and a 4-screws one, are proposed and several configurations (four per architecture) are tested. The overall methodology takes only a few minutes to be performed. The objective of this application is to compare a 3 and a 4-screws solution and to precise the best value of the other parameters to design the optimal solution.

4.2 Results obtained

Tables of failure risks are given on figure 7. Two tables can be seen corresponding to the two architectures tested. A table gives the failure risks for several design criteria and a global failure probability for each configuration: a line corresponds to a configuration and a column corresponds to a criterion with the last one corresponding to the global risk.

The risk are here given by a proportional colour. It can be seen that the more safety solution seems to be the second configuration of the 4-screws architecture. What can be seen too is that the more restrictive criteria are the sixth and the eleventh ones that correspond respectively to the sliding resistance and the fatigue resistance.

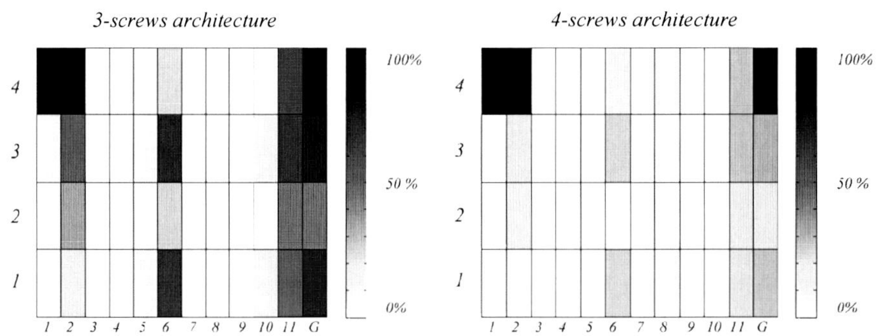

Figure 7: Tables of failure risks for two different architecture suggestions of a mounting (*line* = configuration, *column* = criterion).

These first results help to have an idea of the potential good solutions by comparing the failure risks of the two suggested architectures, and moreover, inside each architecture, several configurations are balanced. Although this information is very useful, it is not enough to take the best decision. Indeed, it would be more interesting to know if the failure risks are concentrated in a limited area or if the failure risks are regularly distributed in the design space. A solution with weak failure risks could effectively be less interesting if the risks are uniformly distributed in the design space, because it will consequently be too difficult to efficiently avoid these unacceptable variants.

4.3 Clusters generated

The following table 2 shows the different clusters found thanks to the algorithms exposed previously. The computation has been made for just one level of quality of the bolts. The table gives the clusters of valid solutions and the clusters of faulty ones for every configuration.

In order to well use these results, the number and the size of every cluster have to be observed. Indeed, the easiest configuration to design is the one that has only valid variants. In this example, the configuration number 3 of the 3-bolts architecture is rather simple to specify because, whereas the majority of the variants are unacceptable, the valid variants are distributed in only two clusters in the design space. Sometimes, like the third configuration of the 4-screws architecture, it is maybe preferable to specify the bad solutions clusters than the good ones because they are less numerous and so easier to take into consideration during the design phase.

This table gives to the designer additional information in the aim of making the best choice. Indeed, this output can be productively used to precise the good configurations: the designer can now take a more hazardous decision if he knows that the solution could be given with specifications in order to avoid the faulty parameters associations and could then certainly decrease the cost.

Table 2: Illustration of the clusters found by the algorithms.

Quality	Architecture.	Configuration	Clusters ok	Statistics ok	Clusters ko	Statistics ko
8.8	3 screws	4	0	-	1	100
		3	2	16 / 6	4	36 / 26 / 12 / 4
		2	3	31 / 17 / 7	3	25 / 17 / 3
		1	3	21 / 6 / 1	3	48 / 18 / 6
	4 screws	4	0	-	1	100
		3	3	48 / 16 / 6	2	24 / 6
		2	3	55 / 24 / 2	2	16 / 3
		1	3	49 / 19 / 5	2	20 / 7

Depending on the failure risk of a configuration, specifications of the good directions or of the faulty ones could be given. For example, the third 4-screws configuration could be specified by the parameters variations of the first valid cluster that contains 48 variants because this specification encircles half the design space and would be so rather easy to follow. But it would be also possible to specify the bad direction that the designer has to avoid, especially for very risky configurations. These specifications could be given as parameter intervals to take into account in the design process and could be also visualized in a skeleton view as shown on figure 8. The skeletons represent the shape of the simplified model used in the methodology; here the bigger valid cluster and the bigger faulty one are drawn. These figures are interesting to give the designer the outline of the design to create and allow having an idea of the global shape of the assembly. For example, in this case it should be better to avoid aligning the two junction beams of the lower bolts because these orientations may lead to failure, but it should be better to orient these beams towards the loading point. In the case of a 3-screws solution that can be seen on figure 9, the main difference between the valid and the unacceptable clusters seem to be the section dimensions of the junction beams.

Finally, in order to make the good choice on this case, it would be interesting to have some comparisons of the costs between the different potential solutions; this aspect will be studied in further works. The final choice will depend on the trade-off that the designer wants to make: he has to choose between an expensive solution with few failure risks and a cheaper solution more difficult to design because subjected to several specifications in order to fulfil the numerous

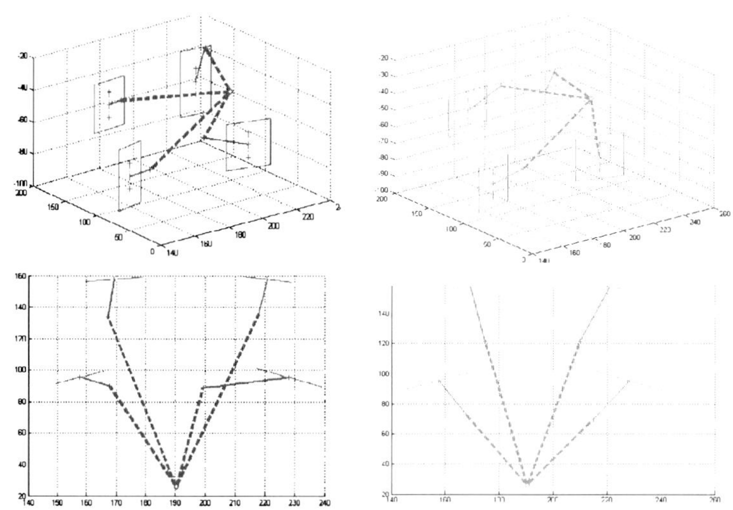

Figure 8: Visualization of the first bad (left) and good (right) clusters in perspective view (top) and in top view (bottom) for the third 4-screws configuration.

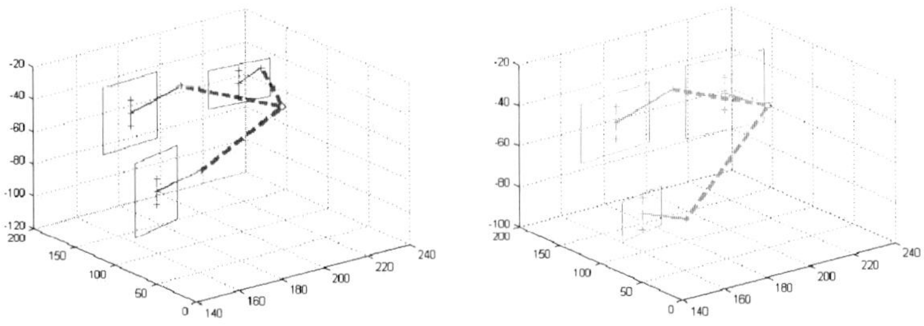

Figure 9: Visualisation of the first bad (left) and good (right) clusters in perspective view for the second 3-screws configuration.

criteria. In this example, it is obvious that the 3-screws solutions are more hazardous than the 4-screws configurations, but depending on the compromise with the cost the designer wants to make, the choice will not be mechanically done for a 4-screws configuration.

5 Conclusion

This paper presents an innovative approach to quickly and efficiently choose the design of bolted joints in the first steps of the design process. The methodology is able to give several potential solutions with associated failure risks estimations and even some specifications in order to efficiently design a risky but cheaper configuration. An industrial application is presented to illustrate the useful results that can be given by the methodology developed.

This work allows giving very useful and relevant information with controlled risks when pre-dimensioning multi-bolted joints in industrial environment. Moreover, the reduced time necessary to obtain such results really contributes to decrease the design process duration without taking uncontrolled risks.

Some further work has to be done to improve the specifications phase, for example by using principal component analysis. Moreover, the overall methodology is tested by giving the results of the computation to designers to check if the outputs of the methodology presented in this paper are well interpreted for them to design the solutions. Then the methodology would be implemented in a design tool in order to be easily used by the designers.

Acknowledgements

This work has been performed within the framework of common researches between PSA Peugeot Citroën, LGMT (Laboratoire de Génie Mécanique de Toulouse) and LTDS (Laboratoire de Tribologie et Dynamique des Systèmes, Lyon).

References

[1] Yannou B., Hamdi A. & Landel E., A strategy for the conceptual modelling of vibro-acoustic performances in the preliminary design stage of a car subframe. *Mécanique & Industries*, vol.4, pp. 365-376, 2003.
[2] Dericquebourg T., Daidié A., Debarle F. & Sartor M., An original global-local approach for predimensioning multi bolted joints. *IDMME-Virtual Concept 2008*, 2008.
[3] VDI Verein Deutscher Ingenieure, Systematic calculation of high duty bolted joints, joints with one cylindrical bolt, VDI 2230, *VDI-Handbuch Konstruktion*, Düsseldorf, 2003.
[4] Alliot J-M., Techniques d'optimisation stochastique appliquées à certains problèmes du trafic aérien, Le algorithmes génétiques (chapitre 1). http://www.recherche.enac.fr/opti/papers/thesis/HABIT/index.html
[5] Ke K., Li W. & Sudjianto A., Algorithmic construction of optimal symmetric Latin hypercube designs. *Journal of statistical Planning and Inference*, 90, pp. 145-159, 2000.
[6] Prescott P., Orthogonal-column Latin hypercube designs with small samples, *Computational Statistics and Data Analysis*, 53, pp. 1191-1200, 2009.

Section 3
Topology optimisation

Some improvements in minimum weight topology optimization with stress constraints

J. París, F. Navarrina, I. Colominas & M. Casteleiro
GMNI—Group of Numerical Methods in Engineering,
Civil Engineering School, Universidade da Coruña,
A Coruña, Spain

Abstract

Topology optimization of continuum structures is a recent field in structural optimization. However, an increasing research activity in this area has been developed since the statement of the very first formulations. These formulations try to obtain the most adequate material distribution that satisfies the imposed structural limitations. The existence or absence of material in each part of the domain is usually defined by using a continuum variable (the relative density) in order to avoid dealing with a discrete optimization problem. This continuum approach of the material properties present important advantages since conventional optimization algorithms can be used. However, numerical models must be considered in order to develop the structural analysis for intermediate values of the relative densities.

In this paper, we present some improvements in a minimum weight approach of the structural topology optimization problem. The main goal of this paper is to present an improved formulation that tries to reach binary 0-1 material distributions by using a continuum approach of the design variables. Furthermore, a perimeter penalization is included in the objective function to simplify the solutions obtained. In addition, some computational aspects are considered in order to reduce the computational effort.

Finally, we compare the solutions obtained by using these formulations in two application examples.

1 Introduction

Topology optimization of continuum structures is a recent field in structural optimization. However, an increasing research activity in this area has been developed

WIT Transactions on The Built Environment, Vol 106, © 2009 WIT Press
www.witpress.com, ISSN 1743-3509 (on-line)
doi:10.2495/OP090071

since the statement of the very first contributions [1, 2]. The traditional statements of this problem try to obtain the optimal distribution of material that maximizes the stiffness of the solution by adequately distributing a predefined amount of material in a predefined domain [1–3]. Thus, the main objective is to determine the parts of the domain where material should exist or not. The existence or absence of material is usually defined by using a continuum variable (the relative density) in order to avoid dealing with a discrete optimization problem. This continuum approach of the material properties present important advantages since conventional optimization algorithms can be used to obtain the optimal material distribution. However, numerical models must be considered to develop the structural analysis for intermediate values of the relative densities. This fact is usually solved by using predefined material microstructure models like SIMP (Solid Isotropic Material with Penalty) [3], for example.

In this paper, we propose a different approach of the topology optimization problem that minimizes the weight of the structure with stress constraints [4–7]. This formulation presents important advantages versus the maximum stiffness approach since the most important instabilities associated to maximum stiffness approaches are avoided. On the other hand, the minimum weight approach with stress constraints requires much larger computing resources than the maximum stiffness approaches. Thus, computational aspects must be addressed in order to reduce the computing effort.

The main objective of this paper is to present an improved formulation that tries to reach binary 0-1 material distributions by using a continuum variable (the relative density). Binary 0-1 material distributions are obtained by using a modified objective function based on the weight of the structure. This approach presents two important advantages since the continuum approximation allows to deal with continuum design variables and binary solutions are finally obtained. Furthermore, a perimeter penalization is included to obtain solutions with a reduced number of trusses.

2 Topology optimization problem

The Topology Optimization problem with stress constraints can be formulated from a generic point of view as

$$
\begin{aligned}
&\text{Find} && \boldsymbol{\rho} = \{\rho_e\} && \\
&\text{Minimize} && F(\boldsymbol{\rho}) = Cost(\boldsymbol{\rho}) && \\
&\text{subject to:} && g_j(\boldsymbol{\rho}) \leq 0 && j = 1, \ldots, m \\
& && 0 < \rho_{min} \leq \rho_e \leq 1, && e = 1, \ldots, N_e
\end{aligned}
\tag{1}
$$

where $\boldsymbol{\rho}$ is the vector of design variables, g_j are the stress constraints, m is the number of constraints considered, N_e is the number of elements of the finite element mesh [7, 8] and ρ_{min} is usually equal to 0.001.

The main issue of this kind of formulations consists in the definition of the most adequate objective function to obtain the benefits expected in the final solution and in the definition of the stress constraints in order to guarantee that the final design satisfies the limitations imposed.

3 Stress constraints

According to that, the authors have proposed in previous publications three different formulations to deal with stress constraints in the topology optimization problem: the local approach, the global approach and the block aggregation of stress constraints [4, 5, 8, 9].

The local approach of the stress constraints states one stress constraint in the central point of each element of the mesh by comparing a reference stress with the maximum elastic limit of the material being used [7, 8, 10, 11]. Thus, if we introduce the required modifications to deal with the singularity phenomena [10, 12], the local stress constraints can be stated as:

$$g_e(\boldsymbol{\rho}) = \left[\widehat{\sigma}\left(\boldsymbol{\sigma}_e^h(\boldsymbol{\rho})\right) - \widehat{\sigma}_{max}\,\varphi_e\right](\rho_e)^q \leq 0, \qquad \varphi_e = 1 - \varepsilon + \frac{\varepsilon}{\rho_e}. \tag{2}$$

The exponent q can be used to deal with real stresses ($q = 0$) or to deal with effective constraints ($q = 1$) [6]. The reference stress $\widehat{\sigma}$ is the Von Mises criterion and $\boldsymbol{\sigma}_e^h$ is the stress tensor of the material in the central point of element e. The relaxation factor ε usually takes the values $(0.001, 0.1)$ [4].

The global approach of the stress constraints is defined to avoid the large computing resources required by the local approach of the stress constraints. Thus, the local stress constraints are aggregated in only one global function by using the Kreisselmeier-Steinhauser approach [4, 7, 13]. Thus, the global constraint can be stated as

$$G_{KS}(\boldsymbol{\rho}) = \frac{1}{\mu}\left[\ln\left(\sum_{e=1}^{N_e} \exp^{\,\mu(\widehat{\sigma}_e^* - 1)}\right) - \ln(N_e)\right] \leq 0, \tag{3}$$

being

$$\widehat{\sigma}_e^* = \frac{\widehat{\sigma}\left(\boldsymbol{\sigma}_e^h(\boldsymbol{\rho})\right)}{\widehat{\sigma}_{max}\,\varphi_e}. \tag{4}$$

The aggregation parameter μ must take the largest value that does not introduce numerical instabilities. These numerical instabilities are usually related to the high non-linearity of the aggregation function [4]. The most appropriate range of values of the aggregation parameter is (20,40). These values have demonstrated to work properly in the application examples solved [4, 5, 8, 9].

The block aggregation of the stress constraints is a more general method that includes both previous formulations. In this approach the domain of the structure is divided in a predefined number of groups of elements such that all the groups contain approximately an equal number of elements. Thus, a global-type constraint can be imposed over the elements of each block of the mesh [5, 8]. Thus,

$$G^b_{KS}(\boldsymbol{\rho}) = \frac{1}{\mu}\left[\ln\left(\sum_{e\in B_b} \exp\,^{\mu(\widehat{\sigma}^*_e - 1)}\right) - \ln(N^b_e)\right] \leq 0 \tag{5}$$

where B_b is the set of elements contained in block b and N^b_e is the number of elements contained in block b.

4 Objective function

The aim of the Topology Optimization problem with stress constraints is to determine the minimum weight structure that supports the applied forces. According to that, the objective function of the optimization problem is the weight of the structure:

$$F(\boldsymbol{\rho}) = \sum_{e=1}^{N_e} \int_{\Omega_e} \rho_e \gamma_{mat} d\Omega \tag{6}$$

where γ_{mat} is the material density, ρ_e is the relative density of element e and ω is the domain occupied by element e.

This objective function introduces an unexpected phenomenon since the material distributions obtained present a large number of elements with intermediate values of relative density, which is an unwanted solution in practice. Thus, a different formulation of this objective function has been proposed in order to penalize the material distributions with intermediate values of the relative density. This formulation has been obtained by following the same idea introduced in the SIMP model of microstructure. This modified objective function can be stated as:

$$F_p(\boldsymbol{\rho}) = \sum_{e=1}^{N_e} \int_{\Omega_e} \Psi_p(\rho_e)\, \gamma_{mat} d\Omega, \qquad \Psi_p(\rho_e) = \rho_e^{\frac{1}{p}}, \qquad p \geq 1 \tag{7}$$

where the exponent p is the penalization factor and takes the values $p \geq 1$ (figure 1 left).

This formulation of the objective function has been widely analyzed in previous publications of the authors and some application examples have been studied [4–8]. However, this formulation does not guarantee a final solution with 0-1 material distribution. This fact can be easily understood if we consider that the derivatives of the objective function are always positive. Thus, a reduction of the design variable always introduces a reduction in the objective function. This fact is the expected one from a physical point of view. However, the solutions obtained with this approach are not 0-1 when we use small values of the relaxation parameter (*e.g.* $\varepsilon = 0.01$). Stress constraints usually impose lower limits of the relative densities. Thus, the lower limits of relative densities imposed by the stress constraints and the positive value of the derivatives of the objective function define an optimal solution that usually presents intermediate values of the relative densities.

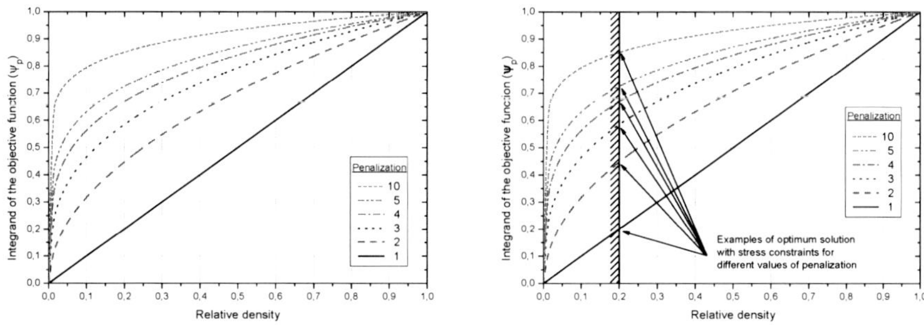

Figure 1: Integrand of the objective function ($\Psi_p(\rho_e)$) for different values of the penalization parameter (p) (left) and examples of optimum design variables when stress constraints are considered (right).

This issue can be explained by analyzing a simplified problem with one design variable and one stress constraint. The optimization algorithm tries to reduce the value of the design variable but the stress constraint avoids the reduction. The optimal solution is obtained for an intermediate value of the relative density although the penalization parameter has been used. This fact can be easily observed in figure 1 (right).

Thus, the proposed penalization of the intermediate relative densities does not guarantee 0-1 solutions when small values of the relaxation parameter are used. The values of the relaxation parameter must be as low as possible in order to avoid removing useful material in the final designs but, on the other hand, it must be large enough to avoid singularity phenomena.

The optimal solutions can be forced to reach 0-1 binary values by modifying the objective function proposed in (7). In this paper, we propose a modified objective function to obtain 0-1 optimal solutions as:

$$F_b(\boldsymbol{\rho}) = \sum_{e=1}^{N_e} \int_{\Omega} \Psi_b(\rho_e)\, \gamma_{mat} d\Omega, \tag{8}$$

$$\Psi_b(\rho_e) = \rho_e^{1/(1+\beta)^2} + \beta \exp^{-\beta\rho_e} \sin(\pi\rho_e), \tag{9}$$

This modification introduces aggressive changes in the mathematical function since the sign of the derivatives becomes negative in part of the range of relative density (see figure 2). This modification produces the expected benefits but it also introduces local minima and other unwanted phenomena. Thus, a whole procedure must be analyzed in order to include this modification in the optimization process.

Most of the initial solutions for the topology optimization problems use the maximum value of the relative density. Thus, if we use the modified objective function proposed in (8) the optimization algorithm will not modify the initial solution since the value of the objective function will rise for a reduction of the design variables. In this case, the optimum solution is the initial distribution and this is not the real

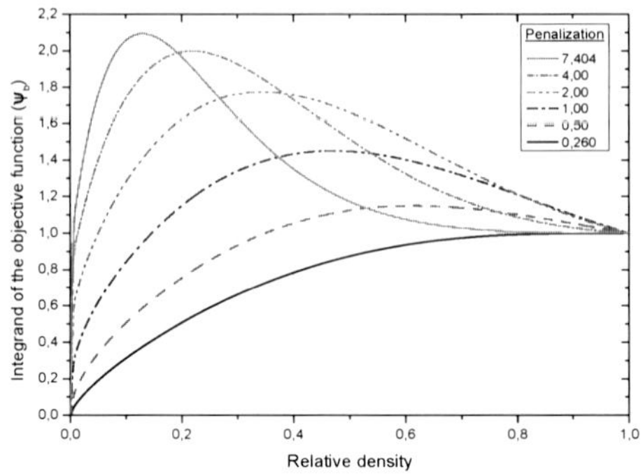

Figure 2: Integrand of the objective function ($\Psi_b(\rho_e)$) (8) for different values of penalization.

optimum solution. Thus, the modified objective function must be specifically used to obtain a final material distribution. The optimization process must be developed in two stages. First stage obtains an optimal material distribution with the objective function proposed in (7) by penalizing the intermediate densities. The second stage starts with the solution obtained in the first stage and incorporates the modified objective function in order to force the elements with intermediate densities to reach 0-1 values. This second stage must be developed by increasing progressively the value of the parameter β proposed in (8).

The values of the parameter β must be in the range $[0.260, 7.404]$ in order to obtain the expected solutions. Values smaller than, approximately, 0.260 introduce positive derivatives when $\rho \rightarrow 1$. Values higher than, approximately, 7.404 also introduces positive derivatives when $\rho \rightarrow 1$.

This formulation of the objective function does not produce minimum weight designs since the modifications introduced forces 0-1 material distributions. These solutions are not the optimal ones in minimum weight terms (when continuous design variables are considered) but they produce important benefits since no elements with intermediate relative densities appear. Thus, the optimal solutions obtained present the lowest cost in practice.

5 Parallel computing

The resulting optimization problem requires the use of efficient algorithms that allow to deal with a large number of highly non-linear stress constraints and a non-linear objective function. In this paper, we have used a Sequential Linear Programming algorithm with Quadratic Line Search proposed in [6, 7, 14].

This algorithm has demonstrated to produce optimal solutions for the optimization problems proposed. However, this algorithm and the sensitivity analysis involved [5, 8, 15] require large computing resources (computing time) when a large number of design variables and stress constraints is imposed. Consequently, it is necessary to propose computational techniques that allow to reduce the compu-

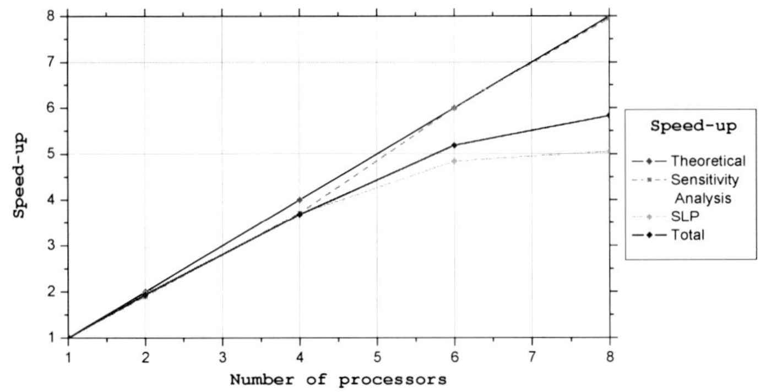

Figure 3: Speed-up obtained for the cantilever beam problem with 7200 design variables by using the local approach of stress constraints in a computer with 4 Intel Xeon 7120M Dual Core processors.

tational effort. According to this idea, parallelization techniques have been implemented in the sensitivity analysis and the optimization algorithm. First order derivatives can be computed independently for each stress constraint. Thus, the parallelization of this procedure is feasible and it produces suitable performance [8]. The Optimization algorithm can be also computed in parallel since the Simplex Algorithm involved develops a large number of matrix operations that can be computed in parallel. The performance of this parallelization is not as effective as in the sensitivity analysis computation since a number of sequential operations must be developed between each matrix modification. This fact can be observed in figure 3. However, the total speed-up of the optimization process allows to reduce the computing time about 6 times when 8 processors are used in a problem with 7200 design variables and constraints.

6 Perimeter penalization

The proposed objective function can be complemented with a perimeter penalization in order to obtain simplified solutions. This perimeter penalization produces optimal solutions with less structural elements (trusses) than the original statement of the objective function (7) and (8). This perimeter approximation is based on the total variation function proposed by Haber *et al.* [16]. Thus, the total variation of the perimeter is defined as:

$$TV(\rho) = \int_{\Omega \setminus \Gamma_J} \|\nabla \rho\| \, d\Omega + \int_{\Gamma_J} |<\rho>| \, d\Gamma_J \tag{10}$$

where $\Omega = \bigcup_\alpha \Omega_\alpha$ and Ω_α is the set of disjointed regions (finite elements) that defines the whole domain Ω. The expression $|<\rho>|$ indicates the absolute difference of relative density between two neighbour disjunct regions Ω_α (finite elements) and Γ_j represents the magnitude of the frontier between two contiguous elements (the length in a 2D problem).

In the topology optimization formulation, the relative density is uniform for each element and, consequently, the first term of equation (10) is null. Thus, the objective function including the perimeter penalization $\widehat{F}(\boldsymbol{\rho})$ can be defined as:

$$\widehat{F}(\boldsymbol{\rho}) = F_p(\boldsymbol{\rho}) + \eta \sum_{\Gamma_J} |< \rho >| L_J \tag{11}$$

where L_J is the length of the frontier between two contiguous elements and η is the weight factor that relates the value of the perimeter to the value of the cost of the structure. This factor is determined by defining a reduced percentage of the ratio obtained by dividing the initial value of the weight by the initial value of the perimeter of the structure. This percentage usually varies from 1 % to 5 % in practical applications. High values of this percentage avoid the generation of trusses in the optimal solution since the minimum perimeter solution corresponds to an equal value of the relative density for all the elements. Thus, great areas with intermediate densities usually appear. On the other hand, low values of this percentage introduce an insignificant effect of the perimeter penalization.

7 Application examples

We present two structural problems frequently analyzed in the topology optimization field. These examples are 2D structures in plane stress. The examples proposed are analyzed by incorporating the perimeter penalization in order to show the effect of this penalization in the optimal solutions obtained. In addition, the effect of the modified objective function proposed in (8) can be also analyzed.

7.1 MBB beam

The first example corresponds to a classic MBB-beam with sliding supports [3]. Only the right half of the structure is analyzed due to symmetry. Figure 4 shows the dimensions of the domain and the position of the external forces (13.3 10^3kN). Self-weight is considered. The domain of the structure is discretized by using $N_e = 120 \times 40 = 4800$ eight-node quadrilateral elements. The material being used is steel with density $\gamma_{mat} = 76500$ kN/m^3, Young's modulus $E = 2.1\ 10^5$ MPa, Poisson's ratio $\nu = 0.3$ and elastic limit $\widehat{\sigma}_{max} = 230$ MPa. The thickness of the structure is 1 m.

Figure 5 (up) shows the solution obtained by using the local approach of the stress constraints and the objective function proposed in (7). Figure 5 (down) shows the optimal solution obtained by using the local approach and the modified objective function proposed in (8) in order to obtain binary 0-1 solutions.

Table 1 shows the most important parameters involved in the optimum design formulation and the value of the final objective function as a percentage of the original weight of the structure. Note that the modified objective function does not

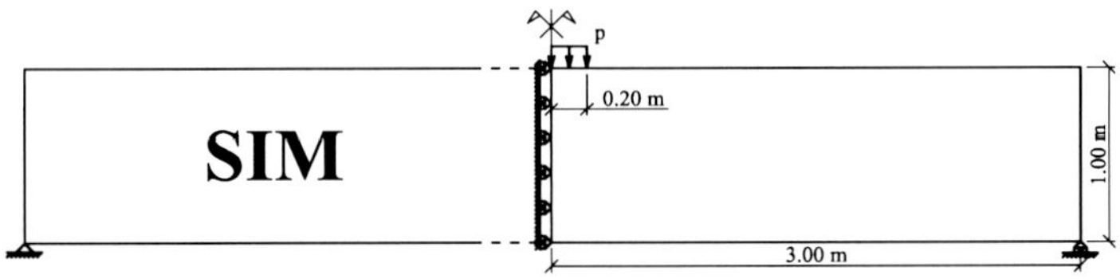

Figure 4: MBB-beam scheme (units in meters).

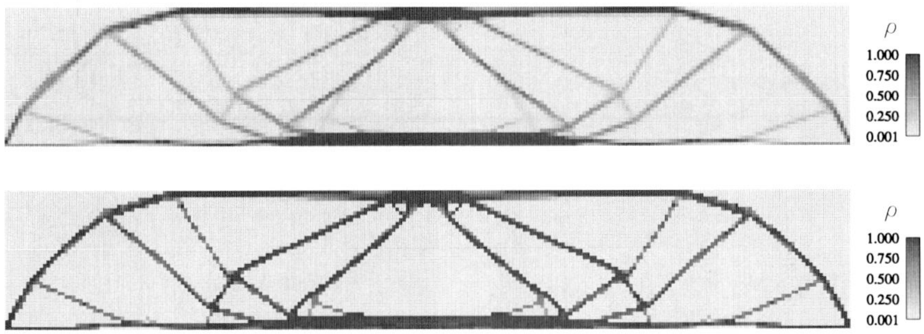

Figure 5: Optimal solution of the MBB-beam problem by using the local approach of the stress constraints (up) and by using the modified objective function (8) (down) ($\varepsilon = 0.01$, $q = 1$, $\eta = 0.03$).

Table 1: Summary of the most important parameters of the MBB-beam problem.

MBB BEAM	Local Approach (Fig. 5 up)	Modif. Objective F. (Fig. 5 down)
Number of elements	4800	4800
Penalization (p, β)	4	7
Final weight/Initial weight	15.43 %	19.60 %

produce minimum weight designs. However, the material distribution is essentially binary. A more accurate solution to minimum weight design with binary distribution can be obtained by using more refined finite element meshes.

7.2 Cantilever beam

The second example corresponds to a cantilever beam with no displacements allowed in the nodes of the left vertical edge. Figure 6 shows the dimensions of the domain and the position of the external load ($2\ 10^3$kN) applied on the border of 7 contiguous elements placed vertically. Self-weight is considered. The domain of the structure is discretized in $N_e = 120 \times 60 = 7200$ eight-node quadrilateral elements. The material being used is steel with density $\gamma_{mat} = 76500$ kN/m^3, Young's modulus $E = 2.1\ 10^5$ MPa, Poisson's ratio $\nu = 0.3$ and elastic limit $\widehat{\sigma}_{max} = 230$ MPa. The thickness of the structure is 0.2 m.

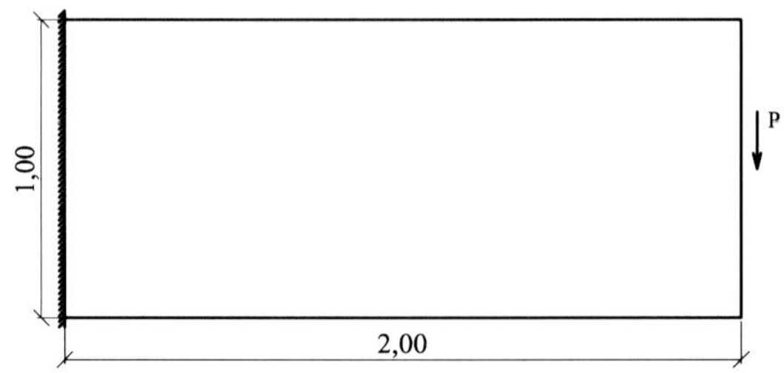

Figure 6: Cantilever beam scheme (units in meters).

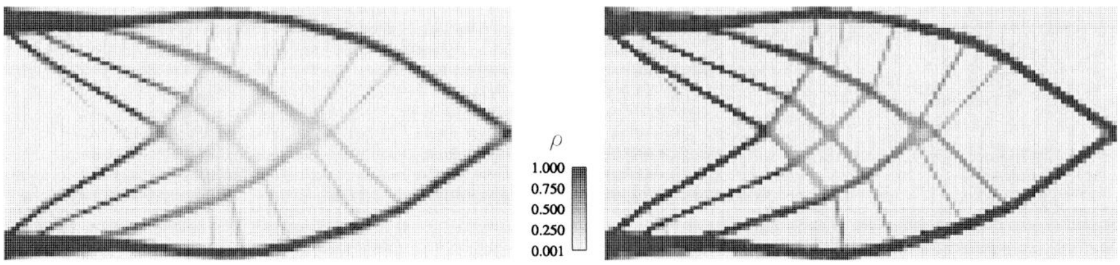

Figure 7: Optimal solution of the cantilever beam problem by using the block aggregation approach (left) and with the modified objective function (8) (right). ($\varepsilon = 0.01$, $\mu = 40$, $\eta = 0$, $N_e^b = 60$).

Table 2: Summary of the most important parameters of the cantilever beam problem.

CANTILEVER BEAM	Block Aggr. Approach (Fig. 7 left)	Modif. Objective F. (Fig. 7 right)
Number of elements	7200	7200
Number of constraints	120	120
Final weight/Initial weight	16.86 %	19.18 %

Figure 7 (left) shows the optimal solution obtained by using the block aggregation of the stress constraints. Figure 7 (right) shows the solution obtained by using the block aggregation of the stress constraints and a final stage with the modified objective function (8). Table 2 presents the most important parameters involved in the optimum design formulation and the value of the final objective function as a percentage of the original weight of the structure.

8 Conclusions

In this paper we present three different formulations to deal with stress constraints in structural topology optimization problems: the local approach of the stress constraints, the global approach of the stress constraints and a more general formulation that defines groups of elements and imposes one global constraint per group.

In addition, we introduce a modified objective function based on the weight of the structure that forces the solution to reach 0-1 values. Thus, elements with intermediate relative densities are completely avoided. This objective function allows to deal with continuum design variables during the optimization process but it produces final binary distributions of material.

We present a perimeter penalization in order to improve the results obtained and we develop the parallelization of the most expensive algorithms in order to reduce the computational effort.

Finally, some application examples are solved in order to verify the validity of the algorithms and formulations proposed in this paper.

Acknowledgements

This work has been partially supported by the *"Ministerio de Educación y Ciencia"* (DPI-2006–15275 & DPI-2007-61214) and the *"Xunta de Galicia"* (PGIDIT03–PXIC118001PN & PGIDIT03–PXIC118002PN).

References

[1] Bendsøe M. P. & Kikuchi N., Generating optimal topologies in structural design using a homogenization method, *Computer Methods in Applied Mechanics and Engineering*, **71**, pp. 197–224, 1988.

[2] Bendsøe M. P., Optimal shape design as a material distribution problem, *Structural Optimization*, **1**, pp. 193–202, 1989.

[3] Bendsøe M. P., *Optimization of structural topology, shape, and material*, Springer-Verlag, Heidelberg, 1995.

[4] París J., *et al.*, Global versus local statement of stress constraints in topology optimization of continuum structures, *Computer Aided Optimum Design of Structures X*, Myrtle Beach (SC), USA, 2007.

[5] París J., *et al.*, Block aggregation of stress constraints in topology optimization of structures, *Computer Aided Optimum Design of Structures X*, Myrtle Beach (SC), USA, 2007.

[6] Navarrina F., *et al.*, Topology optimization of structures: a minimum weight approach with stress constraints, *Advances in Engineering Software*, **36**, pp. 599–606, 2004.

[7] París J., *et al.*, Topology optimization of continuum structures with local and global stress constraints, *Structural and Multidisciplinary Optimization*, DOI: 10.1007/s00158-008-0336-2, 2008.

[8] París J., *Restricciones en tensión y minimización del peso: Una metodología general para la optimización topológica de estructuras* (in Spanish), PhD thesis, University of A Coruña, 2008.

[9] París J., *et al.*, Topology Optimization of continuum structures with local and global stress constraints, *Structural and Multidisciplinary Optimization*, DOI: 10.1007/s00158-008-0336-2, 2008.

[10] Duysinx P. & Bendsøe M. P., Topology optimization of continuum structures with local stress constraints. *International Journal for Numerical Methods in Engineering*, **43**, pp. 1453–1478, 1998.

[11] Duysinx P. & Sigmund O., New Developments in Handling Stress Constraints in Optimal Material Distributions. *7th AIAA/USAF/NASA/ISSMO Symposium on Multidisciplinary Design Optimization*, Saint Louis, Missouri, USA, 1998.

[12] Cheng G. D. & Guo X., *ε-relaxed approach in structural topology optimization*, Structural Optimization, **13**, pp. 258–266, 1997.

[13] Martins J. R. R. A. & Poon N. M. K., On Structural Optimization Using Constraint Aggregation, *Proc. WCSMO6 Congress*, Brazil, 2005.

[14] Navarrina F. *et al.*, An efficient MP algorithm for structural shape optimization problems, *Computer Aided Optimum Design of Structures VII*, WIT Press, Southampton, pp. 247–256, 2001.

[15] Navarrina F. *et al.*, High order shape design sensitivity: A unified approach, *Comp. Meth. Appl. Mech. Eng.*, **1**88, pp. 681–696, 2000.

[16] Haber R. B., *et al.*, A new approach to variable-topology shape design using a constraint on perimeter. *Struct. Opt.*, 12:35–45, 1996.

Three-dimensional cylindrical truss structures: a case study for topological optimization

P. Dewhurst, D. G. Taggart, R. B. Waterman & S. Heinemann
College of Engineering, University of Rhode Island, United States

Abstract

A general minimum-weight cylindrical structural layout for the support of any combination of axial and torsional loading has been developed. The principal intention in this work is to provide a test case for 3-dimensional numerical topological optimization. It is anticipated that the solution may present a challenge, since for the small angular spacing of the truss elements the internal radial force component is always of the order of magnitude of the angular spacing for any arbitrary selected pair of helix families. Moreover for a wide range of solutions slender members are an essential part of the topology.

A novel finite element topology optimization procedure is presented based on the application of Beta probability density and cumulative distribution functions. The procedure utilizes a family of Beta functions which provide a smooth transition from a uniform to a bi-modal density distribution, with constant probability mean to conserve constant mass.

Keywords: Michell structure, topological optimization, beta function.

1 Introduction

Following the initial pioneering work of Michell [1] and a flourish of advances in the 1950's and 1960's (Cox [2]; Hemp [3]; Prager [4]; Chan [5]), the analysis of absolute minimum-weight truss structures entered a period of relative neglect until the 1990's since when it has become a subject of renewed interest; see for example the work by Rozvany et al. [6–8]; Lewinski et al. [9]; Lewinski and Rozvany [10–12]; Graczykowski and Lewinski [13–15]; Dewhurst [16,17]; Dewhurst and Srithongchai [18]; Srithongchai and Dewhurst [19]; Dewhurst et al. [20].

WIT Transactions on The Built Environment, Vol 106, © 2009 WIT Press
www.witpress.com, ISSN 1743-3509 (on-line)
doi:10.2495/OP090081

The theoretical base for all of the above-mentioned body of work was established by Michell [1] who recognized that, in general, truss structures become more efficient as the spacing of the joints or nodes decrease, and so the theoretical optimum is an infinitely-dense but discrete system of tension and compression members. The members can thus curve continuously, supported by a continuum of connections with other intersecting truss members, forming a paradoxical discrete-continuous network. This allowed Michell to conjecture, and then prove, that an optimal truss must follow the orthogonal network of lines of maximum and minimum strain in a constant-magnitude strain field. The problem of determining the positions of the nodes of an optimal truss design was thus reduced to finding an orthogonal network of layout curves on which the nodes should lie.

The renewed interest in Michell structures has resulted from developments of new methods for numerical structural optimization and the consequent need for proven theoretically-optimal test cases. Pioneering work in this area involved the development of specialized variational principle and material volume constraint equations (Bendsoe and Kikuchi [21]; Bendsoe and Sigmund [22]). This work has great fundamental value since it establishes absolute limits on structural weight. The perceived shortcomings are complexity of the method, and the infeasibility of manufacturing resulting geometries, which often comprise porous regions. In response to the former, a much simpler procedure proposed by Xie and Steven [23,24], was based on progressive irreversible removal of material from low strain energy regions. This method unfortunately has slow convergence, and in some cases fails to find known optimal structural geometries (Zhou and Rozvany [25]). A third methodology, based on iterative adjustment of density and stiffness in finite-element iterations recognizes the 'cost' of design realization of low-density porous regions, and assigns disproportionate stiffness reductions to reduced density. This method, described as solid, isotropic microstructure with penalty (SIMP) by Rozvany et al. [26], has been shown to provide good matching of numerical results to known analytical optimal solutions, and to date SIMP is the only method to be offered commercially.

One singular exception, to the finite element based schemes described above, is a method recently proposed by Martinez et al. [27], which expands an initial simplest starting structure by iteratively dividing structural members, and at each step applying an optimization procedure to reposition the structure nodes for least weight. This method is more narrowly focused but displays remarkable convergence to known Michell solutions.

An alternative finite element based topology optimization scheme is described briefly in this paper (Taggart and Dewhurst [28,29]). This procedure is based on an iterative prescribed material redistribution scheme in which the desired material distribution at each iteration is imposed directly. A family of Beta probability density functions is utilized to provide a gradual transition from an initial unimodal material density distribution to a bimodal distribution of fully dense and essentially void regions. The efficiency and validity of the scheme has been demonstrated through a number of 2-D and 3-D test cases for which the optimal topology is known. These test cases include classical minimum weight Michell structures as well as newly derived optimal topologies for 3-D structures. In the present work the results of application of this method to general cylindrical structures is described. The described method is the subject of international patent application number PCT/US2006/062302.

Up to the present time most of the work with both Michell structures and numerical optimization has been two-dimensional. However, a principal focus of the current research project at URI is the investigation of optimal three-dimensional structures, and this gives rise to the need for a non-trivial Michell solution to act as a test case. For this purpose, in the present work, the general case of least-weight cylindrical structures for arbitrary combinations of torsional and axial loading is investigated.

Before developing the general cylindrical structural form, it is useful to review the conditions for optimality of a structure which has equal material properties in tension and compression. Michell's original conditions for optimality require that the structure layout must follow the lines of principal strain field in a strain filed with maximum and minimum strains of equal and constant magnitude. The deformation of the strain filed must be compatible with all of the kinematic behavior of the structure, and of course the structure must be in equilibrium. Michell did not stress the latter requirement since his examples were mainly planar and internally statically determinate. His only non-planar structure is a spherical loxodrome solution which is readily recognized to be in equilibrium due to the symmetry of members and forces surrounding each node.

For the case considered here of axial and torsional loading of cylindrical structures, any pair of families of mutually orthogonal helical strain fields are candidates for optimal structure layout, subject to compatible end conditions. However, for given values of axial and torsional loading, only one member of the candidate pairs of families defines an equilibrium structure; which must therefore be the least weight solution. One final comment is in order before proceeding to derive the structural layouts. Since the strain fields are defined on a principal cylindrical surface, the solutions are only shown to be optimal with respect to alternative structures which do no lie outside of the cylindrical envelope.

2 General cylindrical helical structures

In the case of pure torsion of a cylinder the lines of maximum and minimum strain follow opposite families of 45-degree helices. The truss structure formed by placing nodes, at constant angular increments along these helices, whose connecting members are equal helical chords, thus satisfies all of the requirements of a least-weight Michell structure. A rapid prototyped example of such a structure is shown in Fig. 1.

In this case nodes are positioned at 45-degree angular increments around the cylinder. Pure torsion introduces the same force magnitude into the tension and compression members, and the symmetry of the members surrounding each node ensures that the structure is in equilibrium. In particular the radially inward component of the two tension members at each node is cancelled by the radially outward component of the two compression members. If an axial load is introduced in addition to the torsion load, then tension and compression members will carry different internal forces and radial equilibrium will no longer exist. In this case the structure will be globally unstable at only a fraction of the design load. The general requirement for radial equilibrium will be examined next.

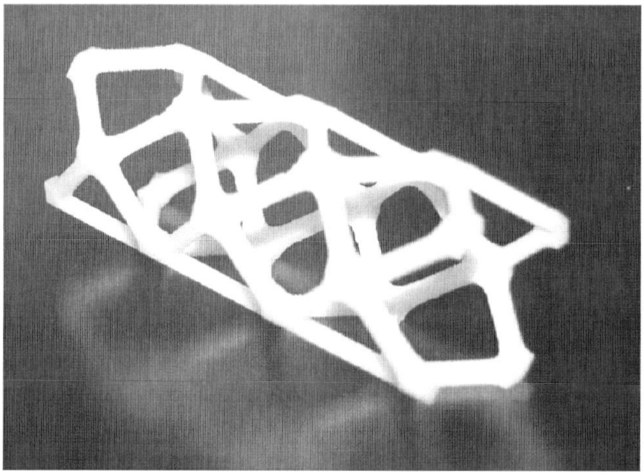

Figure 1: Pure torsion cylindrical structure.

2.1 Condition for radial equilibrium

The combination of axial and torsional loading of course gives rise to principal strain directions which follow two orthogonal families of opposite equi-angular helices. Below we will examine this more general strain field to determine the conditions for equilibrium with arbitrary ratios of axial to torsional loading.

Consider two families of equi-angular helices; left-hand α helices intersecting the axial direction at constant angle γ, and right-hand β helices intersecting the axial direction at constant angle $(\pi/2-\gamma)$. It can be shown that the α helices are represented by the parametric equations

$$x = r\cos(\theta); y = -r\sin(\theta); z = r\cot(\gamma)\theta \tag{1}$$

and the β helices by

$$x = r\cos(\theta'); y = r\sin(\theta'); z = r\tan(\gamma)\theta' \tag{2}$$

where θ and θ' are angles of rotation about the z-axis measured counterclockwise and clockwise respectively from the x-axis. Without loss of generality the α helices will be taken as the layout lines for the tension members and the β helices will lay out the compression ones.

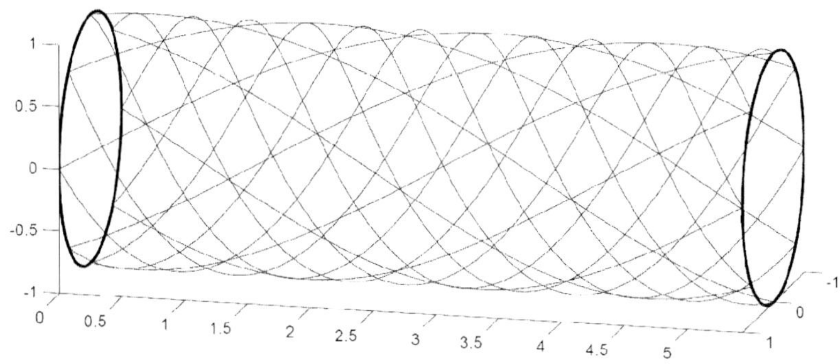

Figure 2: Strain field for axial and torsional loading ($\gamma = \pi/6$).

Figure 2 illustrates the helical network produced by the two helical families for the case $\gamma = \pi/6$.

To establish the relationship between θ and θ' between the nodes of the helical network we can note that they involve equal separation in the axial direction (z) along the two families. Thus from (1) and (2)

$$r\cot(\gamma)\theta = r\tan(\gamma)\theta' \text{ or } \theta' = \cot^2(\gamma)\theta \tag{3}$$

Michell structures may be considered as truss structures with closely spaced nodes so the discrete members follow as closely as possible the curvilinear optimal strain field. For the present case we will consider the nodes to be separated by constant parameter increments $\Delta\theta$ and $\Delta\theta'$ respectively. The members connected an arbitrarily selected node, taken as a local origin, thus have co-ordinate lengths given by Eqs. (1) and (2) as

$$\Delta x = r\Delta\theta^2/2; \Delta y = r\Delta\theta; \Delta z = r\cot(\gamma)\Delta\theta \tag{4}$$

$$\Delta x = r\Delta\theta'^2/2; \Delta y = r\Delta\theta'; \Delta z = r\tan(\gamma)\Delta\theta' \tag{5}$$

for the tension and compression members respectively. Assume the tension members carry internal force f_T and the compression members force f_C. Equations (4) and (5) show that the tension and compression members are orthogonal in the y-z plane to order $\Delta\theta^2$ and so equilibrium in the y and z directions is assured. It therefore only remains to establish equilibrium in the x-direction. From Eqs.(4), the component of tensile force f_T in the x-direction is given by

$$f_{Tx} = f_T\left[\frac{\Delta\theta}{2(1+\cot^2(\gamma))^{1/2}}\right] \tag{6}$$

where the term in brackets is the x-component of the unit vector along the tensile member.

Similarly from Eqs.(5) the x-component of compressive force f_C is

$$f_{Cx} = f_C \left[\frac{\Delta\theta'}{2(1+\tan^2(\gamma))^{1/2}} \right] \tag{7}$$

and substituting for θ' from Eq.(3) gives

$$f_{Cx} = f_C \left[\frac{\cot^2(\gamma)\Delta\theta}{2(1+\tan^2(\gamma))^{1/2}} \right] \tag{8}$$

Substituting Eqs.(6) and (8) into the required condition $f_{Tx} = f_{Cx}$ gives after some trigonometric manipulation

$$f_C = \tan^3(\gamma) f_T \tag{9}$$

2.2 Internal forces in structural members

A necessary step to establish the truss volume is to relate the equilibrium condition, and the axial and torsional loads imposed on the structure, to the internal forces in the structural members. Figure 3 shows the axial and shear force components imposed on a single node from the external loading. Applying axial and circumferential equilibrium conditions on the node, and

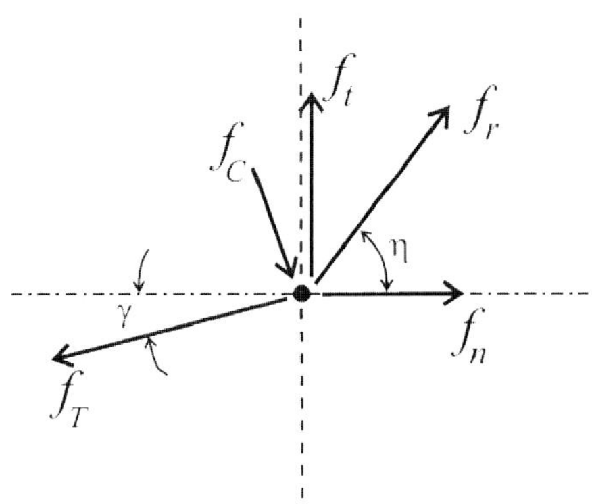

Figure 3: Forces acting on node.

substituting for f_C from radial equilibrium condition (9) gives

$$\begin{aligned} f_n &= f_T(\cos(\gamma) - \sin^4(\gamma)/\cos^3(\gamma)) \\ f_t &= f_T(\sin(\gamma) + \sin^3(\gamma)/\cos^2(\gamma)) \end{aligned} \tag{10}$$

Dividing Eqs.(14) and simplifying leads to the elegant result

$$\gamma = \cot^{-1}((f_n / f_t)/2)/2 = \cot^{-1}(\cot(\eta)/2)/2 \tag{11}$$

where η is the angle of the resultant force; see Fig.3.

This result ranges from $\gamma = \pi/4$ for $f_n=0$ (pure torsion) to $\gamma = 0$ for $f_t=0$ (pure axial loading. For the latter, the structure has become simply a set of axial members at angular spacing $\Delta\theta$ on a pitch circle of radius r. Note also at this point Eq.(9) predicts $f_C = 0$ consistent with the circular column limiting solution.

In terms of the global loading on the structure, the axial force F and torque T are represented by

$$F = 2\pi f_n / \Delta\phi; T = 2\pi r f_t / \Delta\phi \tag{12}$$

where $\Delta\phi = \Delta\theta + \Delta\theta'$ is the angular spacing of nodes around the cylinder.

Thus for any arbitrary combination of axial load F and torque T, the optimal cylindrical structure layout comprises orthogonal families of helices intersecting the axial direction at angles γ and $(\pi/2-\gamma)$ respectively, where

$$\gamma = \cot^{-1}((Fr/T)/2)/2 \tag{13}$$

3 Prescribed material redistribution method

A finite element based topology optimization procedure has been developed in which, through a series of iterative analyses, material is redistributed according to prescribed distributions of nodal densities. This procedure has been named the "prescribed material redistribution" (PMR) method. In this procedure, the desired final mass of the structure is specified at the beginning of the analysis. This material mass is initially distributed uniformly throughout the design domain resulting in a uniform, partially-dense material. All nodes are assigned an initial relative density $\rho_o = V_f / V_D$, where V_f is the final structural volume and V_D is the volume of the partially-dense design domain. This initial distribution can be described by the probability distribution function, f_o, given by $f_o(\rho) = \delta(\rho-\rho_0)$, where δ is the Dirac delta function and ρ is the relative material density ($0 \leq \rho \leq 1$). The corresponding cumulative distribution function, F_o, is given by $F_o(\rho) = H(\rho-\rho_0)$, where H is the Heaviside step function. The desired final material distribution contains two distinct regions of fully dense material ($\rho = 1$) and regions that have zero relative density ($\rho_{min} \ll 1$) with the fully dense regions representing the optimized topology. The final material distribution can be described by the probability distribution function, f_f, given by

$$f_f(\rho) = (1-\rho_o)\,\delta(\rho-\rho_{\min}) + \rho_o\,\delta(\rho-1) \tag{14}$$

and the corresponding final cumulative distribution is given by

$$F_f(\rho) = (1-\rho_o)\,H(\rho-\rho_{\min}) + \rho_o\,H(\rho-1) \tag{15}$$

A gradual transition from the initial distribution to the final distribution can be achieved through the use of the beta function

$$f(\rho) = \beta(\rho, r, s) = \frac{\rho^{r-1}(1-\rho)^{s-1}}{B(r\ \overset{,}{}\ s)} \tag{16}$$

where r and s are adjustable parameters and

$$B(r,s) = \frac{\Gamma(r)\Gamma(s)}{\Gamma(r+s)} \tag{17}$$

where Γ is the gamma function. The corresponding cumulative distribution function, also known as the incomplete beta function, is given by

$$F(\rho) = \beta_{inc}(\rho, r, s) = \frac{1}{B(r,s)} \int_0^{\rho} (\rho')^{r-1} (1-\rho')^{s-1} \, d\rho' \tag{22}$$

A non-dimensional time parameter, t, where $0 \leq t \leq 1$, is introduced and appropriate functions $r(t)$ and $s(t)$ are specified. The functions $r(t)$ and $s(t)$ are selected such that the total mass of material is held constant and a smooth transition from the initial unimodal distribution to the final bimodal distribution is achieved. One such family of distributions are shown in Fig. 4.

At each finite element iteration, nodal densities are assigned based on the sorted nodal strain energies computed from the previous iteration. Nodes with relatively low strain energy are assigned reduced nodal densities and nodes with relatively high strain energy are assigned increased nodal densities. Using a prescribed assignment of nodal densities, the desired progression of density distributions is enforced. In computing the element stiffness matrices, the nodal density field is interpolated to give the Young's modulus, E, at each Gauss point according to the relation $E = E_d \rho$ where E_d is the fully dense Young's modulus. Convergence to the final topology can be achieved in relatively few finite element iterations.

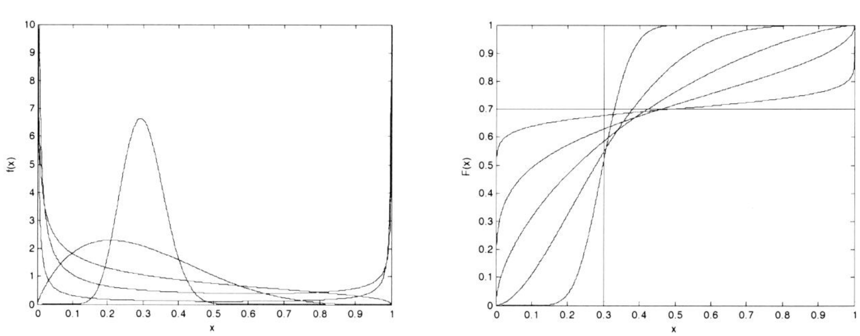

Figure 4: Beta distributions (density and cumulative) showing transition from initial to final distribution for t=0.1, 0.3, 0.5, 0.7, and 0.9.

3.1 Case studies

The numerical procedure has been shown to converge rapidly to a wide range of known two-dimensional optimal truss structures. The case shown in Fig. 5 is for a load at the mid-point of a pinned and a roller support. The selected design domain is shown on the left of the figure and the resulting topology on the right is the well known center fan topology obtained by Michell [1].

For the present problem, the design domain is taken to be a thick walled cylinder subjected to concentrated forces at the ends of the cylinder given by f_n in the axial and f_t in the tangential directions. The optimized results for cases ranging from pure axial loading (f_n=1, f_t=0) to pure torsion (f_n=0, f_t=1) are shown in Fig. 6, together with the effect of decreasing element size.

The helix angle, γ, was estimated from these models and compared to the theoretical optimum given in Equation 17. As shown in Figure 7, the numerical

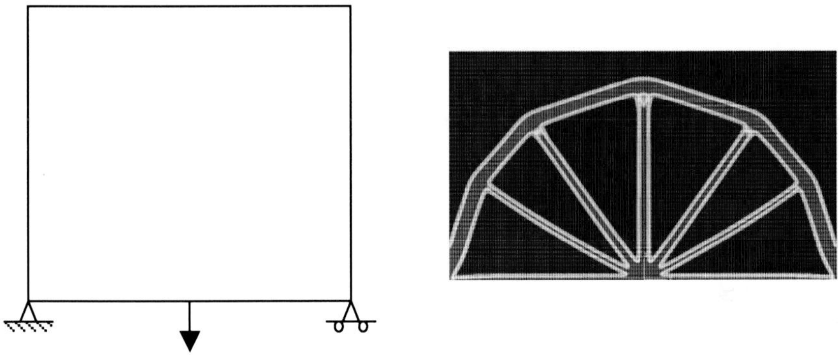

Figure 5: 2-D domain and resulting topology.

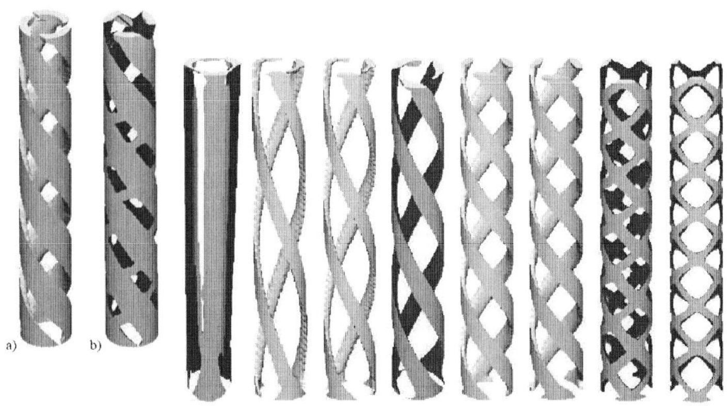

Figure 6: Results for combined axial and torsion loads; including effect of increasing number elements from (a) 5452 to (b) 19584 for (f_n/f_t=1).

scheme accurately predicts the optimum helix angle for all cases. It should be noted that pronounced divergence exists between the helix angles, γ, and the angle of the resultant forces, η, being applied to the boundary. The difficulty of the case, for topology optimization, is that a single set of members tends to propagate from the boundary from the resultant forces, but then follow the optimal helix angle. Producing the complimentary set of supporting helical members is a challenge.

4 Discussion

A generalized minimum-weight cylindrical structural layout for the support of any combination of axial and torsional loading has been developed. The principal intention in this work is to provide a challenging test case for 3-dimensional numerical topological optimization. The case presents two difficulties for topological optimization. The optimal topology is determined by the condition of radial equilibrium. However from Eqs. (6) and (7) it can be seen that any non-equilibrated radial force, resulting from a non-optimal helix angle γ, will only be of the order of magnitude of the node angular separation. The second difficulty stems from the fact that the ratio of the internal forces in the compression and tension helical members is equal to $\tan^3\gamma$. Thus for $\gamma < 25^0$ or $\gamma > 65^0$ one of the helical families will carry less than 10% of the internal load of the other family. Getting both sets of helical members to emerge proved to be a challenge for NMR, as seen in Fig. 6. The authors would be interested in any attempts by interested readers to produce the full family of solutions by SIMP or other methods.

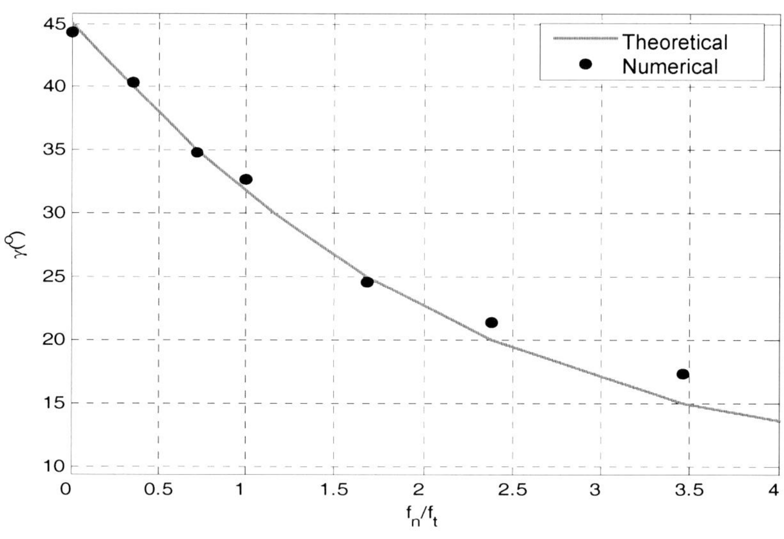

Figure 7: Comparison of theoretical and numerical optimum helix angle γ.

Acknowledgements

This material is based upon work supported by the National Science Foundation under Collaborative Research Grant Nos. 0620752 and by Sandia National Laboratories, Albuquerque, New Mexico.

References

[1] Michell AGM, The limits of economy of material in frame structures. *Phil Mag* **8**, pp. 589-597, 1904.

[2] Cox HL, *The Theory of design.* Aeronautical Research Council 19791, 1958.

[3] Hemp WS, *Theory of the structural design.* The College of Aeronautics, Cranfield. Report No.115, 1958.

[4] Prager W., A problem of optimal design. *Proceedings of the Union of Theoretical and Applied Mechanics*, Warsaw, 1958.

[5] Chan ASL, *The design of Michell optimum structures.* The College of Aeronautics, Cranfield. Report No. 142, 1960

[6] Rozvany GIN, Gollub W, Zhou M, Exact Michell layouts for various combinations of line supports. Part II. *Struct Optim* **14**, pp.138-149, 1997.

[7] Rozvany GIN, Bendsoe MP, Kirsch U, Layout optimization of structures. *Appl Mech Rev* **48**, pp.41-118, 1995.

[8] Rozvany GIN, Querin OM, Logo J, Pomezanski V, Exact analytical theory of topology optimization with some pre-existing members or elements. *Struct Multidisc Optim* **31**, pp.373-377, 2006.

[9] Lewinski T, Zhou M, Rozvany GIN, Extended exact solutions for least-weight truss layouts. Part I. Cantilever with a horizontal axis of symmetry. Part II. Unsymmetric cantilevers. *Int J Mech Sci* **36**, pp.375-419, 1994.

[10] Lewinski T, Rozvany GIN, Exact analytical solutions for some popular benchmark problems in topology optimization II: Three-sided polygonal supports. *Struct Multidisc Optim* **33**, pp.337-349, 2007.

[11] Lewinski T, Rozvany GIN, Exact analytical solutions for some popular benchmark problems in topology optimization III: *L*-shaped domains. *Struct Multidisc Optim* **35**, pp.165-174, 2008.

[12] Lewinski T, Rozvany GIN, Exact analytical solutions for some popular benchmark problems in topology optimization IV: square-shaped line support. *Struct Multidisc Optim* **36**, pp.143-158, 2008.

[13] Graczykowski, C, Lewinski T, Michell cantilevers constructed within trapezoidal domains. Part I. Geometry of Hencky nets. *Struct Multidisc Optim* **32**, pp.347-368, 2006.

[14] Graczykowski, C, Lewinski T, Michell cantilevers constructed within trapezoidal domains. Part III: Force fields. *Struct Multidisc Optim* **33**, pp.1-19, 2007.

[15] Graczykowski, C, Lewinski T, Michell cantilevers constructed within trapezoidal domains. Part IV: Complete exact solutions of selected optimal designs and their approximations by trusses of finite number of joints. *Struct Multidisc Optim* **33**, pp.113-129, 2007.

[16] Dewhurst P, Analytical solutions and numerical procedures for minimum-weight Michell structures. *J Mech Phys Solids* **49**, pp.445–467, 2001.

[17] Dewhurst P (2005) A general optimality criterion for strength and stiffness of dual-material-property structures. *Int. J. Mech. Sci.* **47**, pp.293-302, 2005.

[18] Dewhurst P, Srithongchai S, An investigation of minimum-weight dual material symmetrically loaded wheels and torsion arms. *J of App Mech* **72**, pp.196-202, 2005.

[19] Srithongchai S, Dewhurst P, Comparisons of optimality criteria for minimum-weight dual material structures. *Int J Mech Sci* **45**, pp.1781-1797, 2003.

[20] Dewhurst P, Fang N, Srithongchai S A general boundary approach to the construction of Michell truss structures. *Struct Multidisc Optim* (in press)

[21] Bendsoe MP, Kikuchi N, Generating optimal topologies in structural design using a homogenization method. *Comput Meth Appl Mech Engrg.* **71**, pp.197-224, 1988.

[22] Bendsoe MP, Sigmund O, *Topology optimization: theory, methods and applications*. Springer-Verlag, Berlin, 2004.

[23] Xie YM, Steven GP, A simple evolutionary procedure for structural optimization. *Comput Struct* **49**, pp.885-896, 1993.

[24] Xie YM, Steven GP, *Evolutionary structural optimization.* Springer, London, 1997.

[25] Zhou M and Rozvany GIN, On the validity of ESO type methods in topology optimization, *Struct Multidisc Optim* **21**, pp.80-83, 2000.

[26] Rozvany GIN, Zhou, M, Birker, T, Generalized shape optimization without homogenization. *Struct. Optim.* **4**, pp.250-252 (1992).

[27] Martinez P, Marti P, Querin OM, Growth method for size, topology, and geometry optimization of truss structures. *Struct Multidisc Optim* **33**, pp.13-26, 2007.

[28] Taggart DG, Dewhurst P, A novel topological optimization method. *8th World Congress on Computational Mechanics (WCCM8), Venice , Italy*, 2008.

[29] Taggart DG, Dewhurst P, A finite element based topology optimization scheme based on prescribed material redistribution. *Struct Multidisc Optim* (in review), 2009.

Reciprocating compressor optimum configuration with respect to performance, reliability and commercial conditions

A. Almasi
Rotating Equipment Department, Tecnicas Reunidas S.A., Spain

Abstract

Reciprocating compressors are flexible in order to handle wide capacity and condition swings, offer a very efficient method of compressing almost any gas mixture over a wide range of pressure, can generate a high head independent of density, and have numerous applications and wide power ratings. These make them vital components in various units of industrial plants. In this paper the optimum reciprocating compressor configuration regarding inter-stage pressures, the low suction pressure, non-lubricated cylinder, speed of the machine, capacity control system, compressor valve, lubrication system, piston rod coating, liner material, barring device, pressure drops, rod load, pin reversal, discharge temperature, cylinder coolant system, performance, flow, coupling, flywheel, special tools, condition monitoring (including vibration, thermal and rod drop monitoring), commercial points, delivery and acoustic conditions are presented.
Keywords: reciprocating compressor, optimum configuration, evaluation, performance, reliability.

1 Introduction

Reciprocating compressors are the most common type of compressors found in industrial applications [1–4]. Worldwide installed reciprocating compressor horsepower is approximately three times that of centrifugal compressors and the maintenance costs of reciprocating compressors are approximately three and half times greater than those for centrifugal compressors [3]. The evaluation and purchase of a reciprocating compressor involves an optimization process with respect to the full array of compressor data. Otherwise, equipment may be purchased that is not be suitable over the full range of operating cases, reliability levels, commercial terms or power requirement for the client's needs.

WIT Transactions on The Built Environment, Vol 106, © 2009 WIT Press
www.witpress.com, ISSN 1743-3509 (on-line)
doi:10.2495/OP090091

2 Reciprocating compressor optimum configuration

2.1 Machine design and selection

2.1.1 Inter-stage design pressures

The discharge pressure of each stage is normally protected by pressure relief valves; high pressure discharge switches are seldom seen [2]. Optimum inter-stage pressures can be obtained by the formulation and optimization of performance and investment for the compressor and inter-stage facilities. Some vendors intend to change the inter-stage pressures, with generally higher pressures during the early stages and a lower pressure ratio during the final stages. Vendor-offered inter-stage pressures based on only the compressor without respect to inter-stage facilities are not justified. Inter-stage pressures are going to increase during part load operations and high suction pressure. If this is not acceptable, the additional clearance pocket on the first stage cylinder and the part load operation inhabitation by the controlling logic can be studied. Generally it is a compromise to fix inter-stage facility design pressure based on vendor-recommended PRV set points (around 10–15% higher due to part load operation and suction pressure variations).

2.1.2 Low suction pressure with full flow

Sometimes, due to process requirements, reciprocating compressors are capable of operating at the lowest suction pressure and full design flow at normal discharge pressure. This can have a strong effect on compressor sizing, especially the frame rating and motor power (a power increase of more than 35% for 20% suction below normal). It is optimum to respect this condition as a design point.

2.1.3 Minimum speed-lubricated cylinder

A reliable machine involves low speed (around 350 RPM) and a lubricated cylinder. The optimum piston speed is 3-4.4 m/s. The optimum configuration is horizontal cylinder(s) with a discharge nozzle on the bottom side. For small compressors some vendors intend to deviate the lubricated cylinder or low speed. The optimum option is a lubricated cylinder with the available lowest speed machines. Probably less than 20% of all reciprocating compressors are designed for non-lubricated operation, just because of process demands (oxygen, high pressure air, etc) [2].

2.1.4 Unloader and capacity control system

The stepless capacity control system uses a finger type unloader, is pneumatically actuated, and unloads the suction valve for only a portion of the compression cycle to achieve the adjusted capacity [2]. Finger type unloaders have the potential for damaging the valve sealing elements and require more care for maintenance [2]. Valves and unloaders cause around 44% of unscheduled reciprocating shut downs [5,6] and this selection has a strong effect on reliability [5–8]. For small machines, 100% spill back is the optimum solution, because power is low. For big machines, optimum configuration is the selection of part load steps based on a plug/port unloader and clearance pocket.

2.1.5 Condition monitoring

Condition monitoring [9–13] is particularly cost effective and includes necessary items to identify malfunctions at an early stage (lower maintenance costs and a lower risk of accidents). Optimum vibration monitoring is as follows: 1- Vibration – continuous monitoring (shut down). Velocity transducers are preferred over accelerometers due to a better signal to noise ratio [9]. The optimum configuration is each end of the crankcase about halfway up from the base plate in line with a main bearing [9]. 2- Each cross head accelerometer (alarm). 3- Electric motor vibration (shut down). Optimum temperature monitoring is as follows: 1- high gas discharge temperature – each cylinder (alarm and shutdown); 2- pressure packing case – piston rod temperature (alarm); 3- High cross head pin temperature (alarm); 4- high main and motor bearing temperature (alarm); 5- valve temperature (monitoring); 6- oil temperature out of frame (alarm); 7- high jacket water temperature – each cylinder (alarm). Optimum implementation is properly set trip levels that are just high enough over the normal operating levels to detect mechanical failures, but not so high as to miss the failure prior to catastrophic release [9]. Proximity probes are typically located under the piston rods [9] and used to measure the rod position and determine the wear of the piston and rider bands, malfunction, e.g. a cracked piston rod attachment, a broken crosshead shoe, or even a liquid carryover to a cylinder. The optimum figure is just for the alarm and not for a shut down. Optimum cold run outs and normal condition operating run outs are about 50 microns (2 mils) and on the order of 50 to 150 microns (2 to 6 mils) peak to peak, respectively [9]. All shutdown functions shall be two out of three voting to avoid unnecessary trips. Usually this can be deviated for compressor frame vibration and temperature-related trips.

2.1.6 Valve selection

Cylinder valves are the most critical components of reciprocating compressors and strongly influence the reliability and efficiency [5–8,10]. Valve defects are obviously responsible for most of the unscheduled maintenance events [5–8,10]. Three main valve types are ring type, ported plate and poppet. For big machines (generally low speed and high pressure ratios) and small machines (relatively higher speeds) ring type valves and plate type valves are the optimum choice, respectively. The optimum valve size shall be obtained with respect to efficiency, reliability and performance requirements including minimum clearance volume. Lift is the distance travelled by the valves moving elements. The higher the lift, the higher the valve flow area, the lower the valve pressure drop, the less consumed power, the higher the moving elements impact velocities and the lower the valve durability. An acceptable compromise should be found. Optimum valve spring stiffness is also important. Too much spring stiffness can lead to valve flutter (more compressor power and considerable wear rate) or early closing of the valve (reduced capacity). Too little spring stiffness causes late valve closing and reverse flow (higher velocity, less reliability and reduced capacity). Nonlinear partial differential equations describing the valve differential pressure and the valve element motion (such as [14]) can be used in the optimization process to estimate the optimum valve lift, spring stiffness and gas velocity for each machine and application.

2.1.7 Piston rod coating

The Piston rod seal is the second most important area for the reliability of the reciprocating compressor and the most likely path for potentially hazardous process gas leakage [8]. The packing life could be improved three times by adding the proper tungsten carbide piston rod coating [10], which is the optimum selection.

2.1.8 Cylinder liner material

The cylinder liner is used to provide a renewable wearing surface. The liners made by Ni-Resist cast iron (high Nickel content) are not recommended due to problems such as permanent distortions. The optimum selection is grey cast iron [8], except in very high pressure or extremely high corrosive applications.

2.1.9 Passive vibration reduction system

Sometimes an odd number of cylinders is avoidable. In this case a dummy crosshead shall be used to reduce vibration. The spring-mass-spring system shall also be studied for passive force counter balance and a further reduction in vibration, where the dummy crosshead is, on the one hand, flexibly attached to a movable piston assembly and on the other hand, to the stationary compressor casing using auxiliary mechanical springs.

2.1.10 Future expansion

Future expansion planning can save money and time if the process changes (capacity increase, molecular weight increase due to catalyst change, etc) are foreseen [12]. The optimum selection is sizing cylinders for economical operation at the present rate; the frame can be sized for future applications. When the future conditions become a reality, the cylinders can be changed while keeping the same frame. Generally it is optimum to over size the journal diameter to include margins for future development, thus ensuring that crankshaft size would never become the first important limitation of the design [15].

2.2 Commercial optimum conditions

2.2.1 Commercial conditions

It is absolutely necessary to receive at least three proposals and have, as a minimum, two technically accepted machines. It is completely justifiable to extend the proposal deadline, extend the clarification time, accept an optimum configuration, allow reasonable deviations and attend extensive clarification meetings to have at least two clarified and technically accepted proposals.

2.2.2 Delivery

Small and medium machines are delivered fully fabricated as one skid mounted package. For very big machines, the optimum situation is to deliver the machine prefabricated (including the crankcase, distance pieces, etc) while the cylinders are dismantled. Assembled cylinders are delivered to the site separately and installed. The vendor is advised to offer all site supervision work for cylinder installation as a closed price.

2.3 Auxiliaries and accessories

2.3.1 Lubrication system

API 614 is typically applied only to reciprocating compressor trains involving a large turbine driver and gear unit [11]. The optimum oil system includes two oil pumps (for special purpose machines as per API 676), both sized at least 20% over that required for current operation (two motor driven identical pumps with run down tanks, or a well known crank shaft driven main oil pump; supplying UPS power for one pump is not acceptable alternative), a dual removable bundle shell and tube oil coolers (TEMA C) and double oil filters with a removable element and stainless steel piping.

2.3.2 Coupling

Because of special design of the reciprocating machine, the potential exists for torsional resonance and torsional fatigue failure [2]. Coupling is the best available option for modification to tune the system. The available coupling configurations are: 1- high torsional stiffness coupling (this is the optimum if allowed by torsional analysis); 2- flexible coupling (more elasticity and damping and more maintenance); 3- a direct forged flanged rigid connection (no coupling), with a single bearing motor. Coupling for big machines shall be as per API 671.

2.3.3 Coolant system

Liquids should never form inside the cylinder [10–12]. Liquid contributes to poor reliability, can cause high impact velocities, can lead to the stressing of valve moving elements (slugging) and reduce the lubrication effectiveness. For any application, a good-sized suction drum with a drain provision shall be in order [12]. It may be a part of the pulsation control, if properly done. The cylinder cooling system must be monitored and controlled. The coolant inlet temperature should be between 6 °C and 16 °C above the inlet gas temperature [12]. For exotic gases or operations near critical areas, much care also needs to be taken to make sure the thermodynamic model is proper.

2.3.4 Pulsation control

The optimum technique is to dissipate less energy rather than rely on special solutions, such as orifices to control pulsation levels [16]. Acoustic reviews shall be performed for the design and guarantee points, as well as all other operating cases and the combinations of pressures, speeds and load steps [13]. Pulsation can also alter the timing of the valve motion and decrease efficiency and reliability [5]. The preliminary optimum value for the maximum allowable peak to peak is recommended as Eq. (1), in SI units, where P_1 and P_L are a percentage of the average absolute pressure (%) and the line pressure (Bara), respectively.

$$P_1(\%) = \frac{6}{(P_L)^{1/3}} \tag{1}$$

2.3.5 Pulsation shaking forces

The reduction of pressure pulsation can be accompanied by an increase in shaking forces (or unbalanced forces) [17]. It illustrates that shaking forces shall be determined and controlled and piping and vessels properly supported. The margin of separation between the mechanical natural frequency (MNF) of the system (including piping and bottles) and excitation frequency is 20% and the MNF shall be greater than 2.4 times the maximum run speed [11,18]. If these limits are not met, the force response (including stress analysis) is required. The cylinder gas forces (also called frame stretch or cylinder stretch force) can be a significant source of excitation (they can cause high frequency vibration on the bottles and piping close to the compressor) and lead to excessive pulsation bottle vibration, even if the pulsation shaking forces meet limits. Flow induced pulsation is rarely seen [17]. API 618 Design Approach 3 and less rigorous analysis to control pulsation and shaking (unbalanced) force levels and avoiding mechanical resonance can result in an optimized design [17]. The pulsation and vibration analysis report includes Time Domain (TD) and Frequency Domain (FD) simulations, Time Domain (TD) plots of key forces and pressure pulsation, dynamic pressure drop, models including mounting details (mounting plate, bolts, localized skid, etc) and shell flexibility (nozzle connection flexibility), calculated cylinder stretch forces, the mode shape of bottles and piping and compressor stiffness assumption (compressor frame modelled as a flexible support) [18].

2.3.6 Coolers

The inter stage cooler and the after cooler should be sized carefully. An undersized cooler can cause excessive pressure drop and power loss. Pulsation and shaking force studies are necessary in order to avoid a vibration problem in the cooler. An increased cooler cross section area to decrease pressure drop can cause a significant increase of shaking force and cooler vibration. Secondary volumes may be studied to reduce this vibration, however, in some cases this solution cannot reduce vibration, and modifications to the recycle line are required in order to significantly lower shaking [17].

2.3.7 Torsional analysis

Typically, for a reciprocating compressor, the lateral natural frequencies will be positioned well above significant torsional natural frequencies, so lateral critical studies are not required. A stress analysis should be performed if the torsional excitation falls close to the torsional natural frequency, in order to ensure that the resonance will not be harmful for the system [18]. The torsional vibration analysis report should include data used in the mass elastic system, a display of force versus speed, torsional critical speeds, deflection (mode shape diagram), worst case design, upset condition results (such as valve failure, start up, short circuit, electrical network faults, etc) and how the input data variance will affect the results (sensitivity analysis) [18]. Continuous operation at torsional resonance should be avoided. Changing the load sequence could help to reduce torsional vibration. Full load shutdowns should be avoided. Torsional vibration should be measured and verified during the performance test. Synchronous motors or systems started on a frequency basis need more care (they definitely need a transient torsional start up analysis). The electric machine (driver) shaft diameter should be equal to or greater than the reciprocating compressor crankshaft diameter.

2.3.8 Flywheel and irregularity

For all reciprocating compressors, the flywheel is mandatory to regulate variable reciprocating torques. The irregularity degree for mechanical component reliable operation is a maximum of 2%. This value is the minimum requirement for all compressors. Generally, in accordance with specific requirements for drivers (especially current pulsation for electric motors), torsional vibration results, etc, a lower irregularity value is required. It is recommended to obtain 1% for special purpose units. The optimum irregularity value is between 1-1.5%.

2.3.9 Dynamic package analysis

The dynamic package analysis shall include modelling and simulation of the foundation at the same time. The accuracy of this analysis is strongly influenced by the design of the foundation, especially for pile installation [18]. It is even more important for packages mounted on offshore platforms, Floating Production Storage Offloading Vessels (FPSO), modules mounted on steel structures, new and unproven skid design and where the local soil conditions are suspect. A skid lifting study (including lifting lug details and a calculations review) is necessary. A transit study and environmental loading analysis are also recommended [18].

2.3.10 Barring device

When a compressor is stopped for an extended time, it should be turned around a quarter turn every week by a barring device. A manual barring device can be used for a small compressor. A pneumatic barring device should be used for a compressor rated over 750 KW, which is without an area classification problem or intermittent power availability and is preferred technically [11].

2.3.11 Special tools

The optimum check list for special tools [11] for big machines is as follows: 1- bearing extractor; 2- piston extractor; 3- valve extractor; 4- piston fit up tool; 5- hydraulic tightening system; 6- crosshead assembling tool; 7- special lifting tools; 8- partition plate assembling tools; 9- mandrels for wear bands. For special tools, toolboxes are required and they should be delivered with the main machines, in separate and tagged boxes.

2.3.12 Package general arrangement

Layout is often complex and compromise must be made between the support requirements, process requirements, vibration/pulsation conditions, access and maintenance. The optimum configuration is to install local panels near to the compressor package (around 200 mm away from package), but on a separate skid (frame), which is installed on a foundation in order to avoid vibration damage. For maintenance, consider spool removal and avoid support removal. Access is required in front of the cylinder for cylinder piston dismantling and the non-drive end of the compressor. Attention should be paid to the minimum elevation requirement of the pulsation bottle suction flange to keep suction line "No Pocket".

2.3.13 Piping analysis

The thermal piping design often requires that flexibility be added to the system, which is counter to requirement for more support and increased stiffness to meet vibration design. These analyses should be conducted by the same party in order to optimize design iteration and result in an overall optimized system. Piping thermal analysis is necessary [18], especially when the coolers are off-skid, there are multiple compressors on a common header, there is an extremely cold ambient temperature, or the operation is over a very wide range of conditions.

2.3.14 Cylinder lubrication system

Using the proper type of lubricant, as well as establishing the proper lubricant rates for the cylinder and packing, can be most important for machine reliability [10]. The life of the compressor valves, piston rings, rider bands, and pressure packing can all be significantly affected by the type and quality of lubrication used. Too much or the wrong type of lubrication can increase the effects of valve stiction (viscous adhesion) and result in a reduction in reliability. A reliable lubrication system with an optimum type and rate of lubrication should be selected [10].

2.4 Performance

2.4.1 Expected pressure drops

The preliminary optimum pressure drop values are pulsation dampeners: 1% pressure and intercooler: 0,70 bar. The use of orifice plates, especially on high-speed single-acts, can contribute to significant pressure drops [13].

2.4.2 Rod load and pin reversal details

Maximum rod load is recommended to be less than 80% of the allowable rod load. The duration and peak magnitude load of the rod reversal should not be less than 15° of the crank angle and 3% of the actual combined load [11] in the opposite direction, respectively, and should be checked for all possible operating cases [13,19] (especially low suction and part load).

2.4.3 Highest expected discharge temperature

High discharge temperatures cause problems with lubrication cooking and valve deterioration [12] and should be reviewed at least for average and maximum suction temperatures [13]. The maximum predicted discharge temperature [1,2,4,11] should not exceed 150°C, and should not exceed 135°C for hydrogen rich services (MW of 12 or less). Gas discharge temperatures of less than 118°C tend to extend life of wearing parts [10]. This is an optimum figure.

2.4.4 Performance curves

The required performance curves [13] are: 1- suction pressure versus load; 2- suction pressure versus flow; 3- discharge pressure versus load; 4- discharge pressure versus flow; 5-suction pressure versus discharge pressure, per load step. Performance curves are used to safely control the unit across its defined operating range [13].

2.4.5 Flow details

Flow details are flow curves from a unit's minimum achievable flow rate to its maximum achievable flow rate, in specified increments [13]. An alternative is flow versus discharge pressure plots of specific suction pressures (more compact and common when suction pressure variation is limited).

2.4.6 Load step curvature

A review of the steepness of the load curves can quickly identify which load steps (and where) are quite steep in nature, and thus small changes in pressure can have significant changes in load and flow. Often, steep load curves may indicate improper sizing of cylinders and units with steep load step curves can also prove difficult to automate and tune [13].

2.4.7 Machine shop run test

The shop mechanical run test [11] is the first test after the manufacture of a machine and the last test before delivery. The shop test results, including vibration, may seem to have limited usefulness because the supporting structure and operating condition of the machine are different with final site installation; however this test is a unique opportunity to find defects in the design and manufacturing phases while the machine is still in the fabrication shop. Generally, larger power per compressor throw leads to higher dynamic forces and 370 KW or more per throw is a high-risk machine [20]. It is justified and optimum to record and analyze shop run test measurements (a Kalman filter can be used to optimally evaluate machine dynamic characteristics based on measured data).

3 Conclusion

The optimum configuration of reciprocating compressors, with regard to component design and selection, commercial points, auxiliary and accessories, performance and reliability, are addressed in this paper.

References

[1] Heinz P. Bloch, Compressor and Modern Process Application, John Wiley and Sons, 2006.
[2] Heinz P. Bloch, A Practical Guide To Compressor Technology, Second Edition, John Wiley and Sons, 2006.
[3] W. A. Griffith, E. B. Flanagan, Online Continuous Monitoring of Mechanical Condition and Performance For Critical Reciprocating Compressors, Proceeding of the 30th Turbo-machinery Symposium, Texas A&M University, Houston, TX, 2001.
[4] Heinz P. Bloch and John J. Hoefner, Reciprocating Compressors Operation & Maintenance, Gulf Publishing Company, 1996.
[5] S. Foreman, Compressor Valves and Unloaders for Reciprocating Compressors – An OEM's Perspective, Dresser-Rand Technology Paper, http://www.dresser-rand.com/e-tech/recip.asp.

[6] Steve Chaykosky, Resolution of a Compressor Valve Failure: A Case Study, Dresser-Rand Technology Report, Dresser-Rand Technology Paper, http://www.dresser-rand.com/e-tech/recip.asp.

[7] Massimo Schiavone, Evaluation of The Flow Coefficient of Cylinder Valves, Compressor Tech Two, pp. 48–50, April 2007.

[8] Robin S. Wilson, Reciprocating Compressor: Reliability Improvement Focusing on Compressor Valves, Piston and Sealing Technology, Compressor Optimization Conference, Aberdeen, 30–31 January 2007.

[9] Steven M. Schultheis, Charles A. Lickteig, Robert Parchewsky, Reciprocating Compressor Condition Monitoring, Proceeding of the Thirty Sixth Turbomachinery, pp 107–113, 2007.

[10] Stephen M. Leonard, Increasing the Increase Reliability of Reciprocating Hydrogen Compressors, Hydrocarbon Processing, pp. 67-74, January 1996.

[11] Reciprocating Compressor for Petroleum, Chemical and Gas Service Industries, API 618 5th edition, December 2007.

[12] Royce N. Brown, Compressors Selection and Sizing, Third Edition, Gulf Publishing, 2005.

[13] Dwayne Hickman, Specifying Required Performance When Purchasing Reciprocating Compressor – Part I, II, Compressor Tech Two, August – September–October 2007.

[14] Enzo Giacomelli, Fabio Falciani, Guido Volterrani, Riccardo Fani, Leonardo Galli, Simulation of Cylinder Valves For Reciprocating Compressors, Proceeding of ESDA 2006, 8th Biennial ASME Conference on Engineering Systems Design and Analysis, July 4–7, 2006, Torino, Italy.

[15] Ian Cameron, Thomassen Prescience Pays Dividends, Compressor Tech Two, pp 12–13, November 2007.

[16] A. Eijk, J.P.M. Smeulers, L.E. Blodgett, A.J. Smalley, Improvements And Extensive to API 618 Related To Pulsation And Mechanical Response Studies, The Resip – A State of Art Compressor, European Forum for Reciprocating Compressor, Dresden, 4–5 Nov. 1999.

[17] Brain C. Howes, Shelley D. Greenfield, Guideline in Pulsation Studies for Reciprocating Compressors, Proceeding of IPC 02, 4th International Pipeline Conference, Calgary, Alberta, Canada, Sep. 29–Oct.3, 2002.

[18] Shelley Greenfeld and Kelly Eberle, New API Standard 618 (5 TH ED.) and its Impact on Reciprocating Compressor Package Design – Part I, II and III, Compressor Tech Two, June–July–August 2008.

[19] Alberto Guilherme Fagundes, Nelmo Furtado Fernandes, Jose Eduardo Caux, On-line Monitoring of Reciprocating Compressors, NPRA Maintenance Conference, San Antonio, May 25–28, 2004.

[20] Kelly Eberle and Chris Harper, Dynamic Analysis Of Reciprocating Compressors On FPSO Topside Modules – Part I & II, Compressor Tech Two, pp 10–16 and pp 42–48, April & May 2007.

Section 4
Design optimisation in materials, construction and bridge engineering

Cable optimization of a long span cable stayed bridge in La Coruña (Spain)

A. Baldomir & S. Hernández
School of Civil Engineering, University of Coruña, La Coruña, Spain

Abstract

This document describes an optimization problem of cable cross section of a cable stayed bridge considering constraints of cable stress and deck displacement. Since the bridge is still in study phase, the geometry and the mechanical characteristics are subjected to changes. In order to avoid creating different structural models, a program was produced to construct a model from geometrical and mechanical data and resolve optimization problem. At the end of the document, two examples are presented to show how this program works.
Keywords: *optimization, cable stayed bridges, civil engineering.*

1 Introduction

The optimization techniques are not commonly employed in professional life in civil engineering, specifically in the area of structural calculation, and the majority of consulting firms do not apply these techniques to their designs. Nevertheless, in the design of great structures, these techniques are gaining more importance due to their impact on reduction in material cost.

The presented work comes from the general study of traffic systems around A Coruña, in particular, the connection of the littoral area of Oleiros and Sada to the city. After analyzing traffic in different roads, their intensity, and future projects of hub connections, it was concluded that it would be necessary to construct a new road to be able to efficiently relieve heavy traffic.

The new connection requires a construction of a bridge over the Coruña estuary, which is the most important part of the road. Then a study was conducted on the bridge typology that fits the best to the technical, environmental, and aesthetic conditions of the area. After considering different proposals, the typology of cable stayed bridge was found to be the most adequate.

WIT Transactions on The Built Environment, Vol 106, © 2009 WIT Press
www.witpress.com, ISSN 1743-3509 (on-line)
doi:10.2495/OP090101

Cable stayed bridges are used more and more to overcome long spans and there already exists one with one kilometer of span length [1]. The reasons to use such typology are for its good structural functionality, visual lightness, great aesthetic component, and its low impact on the environment, which was the key issue at the location of the bridge over the Coruña estuary.

The work presented in this document takes part in the project from this point, and optimization of the cable cross section of the cable stayed bridge is developed.

2 Bridge description and structural model

After the study the road solution considers the construction of a cable stayed bridge of 1198 m of total length which is divided into two lateral spans of 270 m and the main span of 658 m.

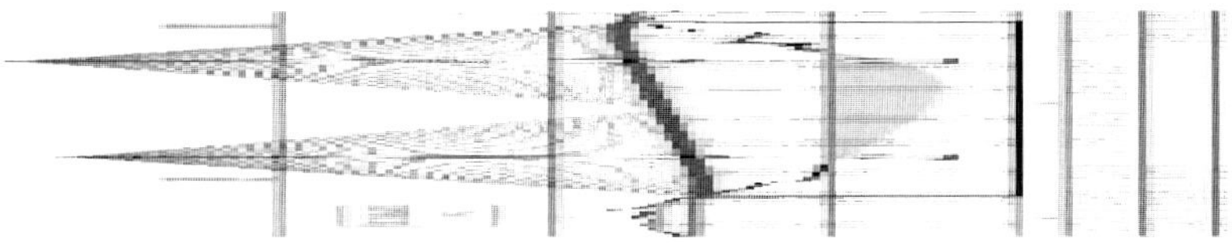

Figure 1: Side view of the bridge.

The main elements that compose the bridge are:

- *Deck*: made of steel with an aerodynamic section of 3 m of height and 34 meter of width, which permits a configuration of 3 lanes in each direction. The difference in height between two end points is 15 m.
- *Towers:* the bridge has two towers: each fixed in the foundation with two twin concrete piers separated by the deck width as shown in Figure 3. From the deck two steel masts come out in the form of λ united by transversal bracing and their upper part continue as a single vertical mast. The cables are anchored to this part of the tower. Moreover, there are other transversal bracing at the lowest part of the vertical mast to avoid buckling problem.
- *Cables*: the bridge deck is sustained by 80 pair of cables that connect to the towers. The cable configuration is hybrid of harp and fan.

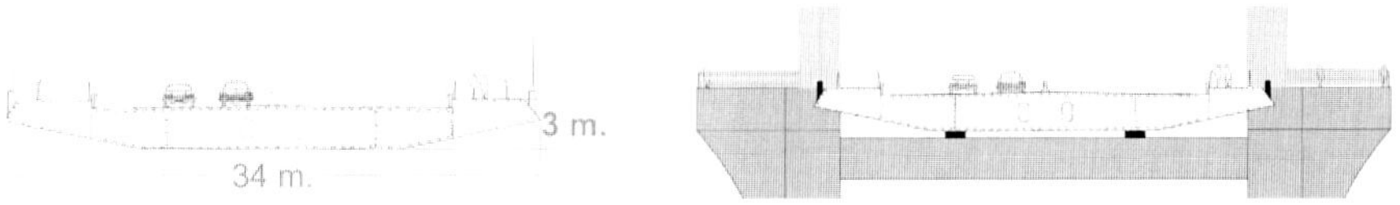

Figure 2: Cross section of the deck.

The optimization process of the cables requires successive structural analysis of the bridge. A finite element model with beam elements was generated in ABAQUS code [2]. As will be seen later, this model is not fixed, but parameterized for avoiding to remodel due to eventual changes in the bridge design.

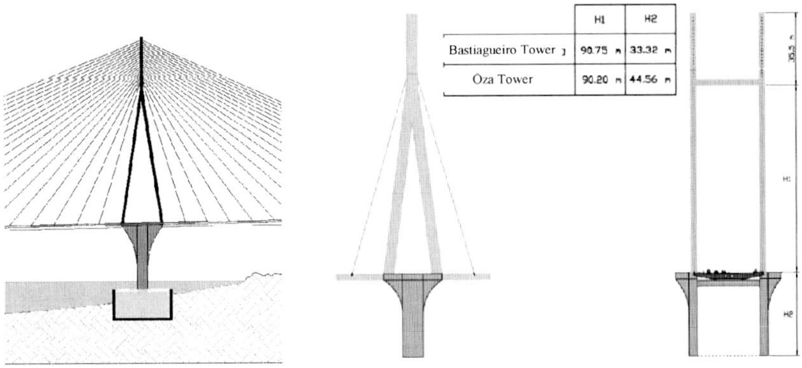

Figure 3: Tower details.

Figure 4: Visualization of the new bridge over the Coruña estuary.

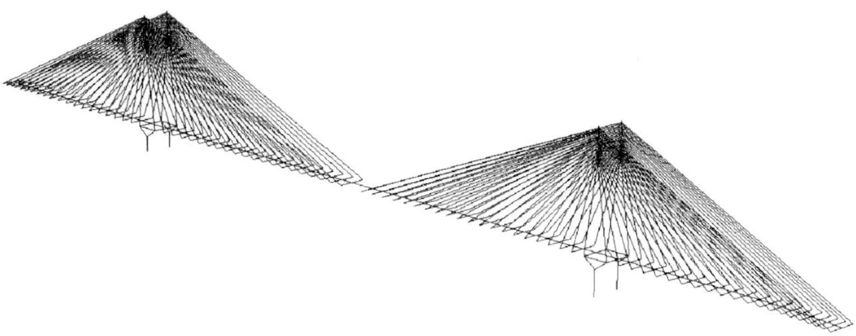

Figure 5: Structural model of the bridge.

To model the structure, 3D beam element, B31 is used with its rotations released at the beam ends for the cables. The material utilized is steel for the deck and the masts of the towers, and concrete for the lower part of the towers. A linear elastic model is considered with isotropic character for Young's module and Poisson's ratio.

3 Formulation of optimization problem

The objective of this work is: to create a program that is able to generate a generic finite element model of a cable stayed bridge (according to the structural scheme proposed in the previous section), to be calculated in ABAQUS, and from the data obtained in this calculation, to perform optimization process on the cable cross sectional area.

The optimization problem is defined by the following elements [3–4]:

- ***Design variables***

The design variables are each of the cable cross sectional areas. The number of design variable is reduced to half of the number of cables due to symmetry about the longitudinal plane; however, it is not symmetrical about the perpendicular plane. For better performance of the optimization process, the inverse of the areas are used to linearize the stress constraints of the cables.

$$x_i = 1/A_i \qquad i = 1,\ldots,n$$

From now on, we call "n" total number of design variables.

- ***Design constraints***

Three types of design constraints were taken into account:

1. Stress constraints of the cables

$$\sigma^k(x_i) \leq \sigma_M \qquad i = 1,\ldots,n \quad k=1,\ldots,LC$$

where:

$\sigma^k(x_i)$: normal tensile stress in the cable, "i" for the load case, "k"

σ_M : maximum allowable normal tensile stress in the cables

LC : total number of load cases

2. Displacement constraints of the deck

$$\left|w^k(x_i)\right| \leq w_{\max} \quad i = 2,\ldots,n-1 \quad k=1,\ldots,LC$$

where:

$w^k(x_i)$: deck displacement at the cable position "i" for the load case "k".

$w_{\max}$: maximum allowable displacement (positive value)

3. Displacement constraints of towers in the longitudinal direction

$$\left| u^{k}{}_{Tower_j} \right| \leq u_{\max} \quad j = 1,2 \quad k=1,\ldots,LC$$

$u^{k}{}_{Tower_j}$: displacement in "x" direction at the top of the tower "j" for the load case "k".

$u_{\max}$: maximum displacement in "x" direction at the top of the tower (positive value).

- ***Objective function F(X)***

The objective of optimization in this case is to reduce the quantity of steel for the cables. Thus the objective function is expressed as:

$$\min \; F = 2 \cdot \sum_{i=1}^{n} L_i \cdot \frac{1}{x_i}$$

where:

L_i : cable length "i"

As we can see, this is an optimization problem with "n" design variables, 2n x LC inequality constraints, and an objective function.

4 Program description

The resolution of the problem formulated previously is carried out by a main program created in MATLAB [5] and the use of software ABAQUS for the structural calculation. This program basically consists of three parts:

1. Generation of finite element models.
2. Calculation of initial stresses in the cables for the self weight.
3. Optimization of the cable sections according to the established load hypotheses.

4.1 Generation of structural model

In this part of the program, we assign dimensions, mechanical characteristics, and discretization for each of the elements that compose the structural model.

This part of program consists of three subroutines, "inputtablero", "inputtorres" and "inputcables" which are in charge of writing Abaqus code according to the data provided initially.

Initial values of design variables are assigned to each cable section in the program.

4.2 Initial tension force in the cable

The construction of the bridge is executed by balanced cantilever from each of the towers. The constraint to be satisfied in each phase of the process is the displacement of the deck where cables are attached should be null. Therefore, just before placing the last center deck section, the bridge should have zero displacement, and after the last deck section is placed, static analysis calculation should show only slight axial forces in such deck section. If the calculation of initial tension with the complete model was carried out, this last deck section would have axial forces above the actual ones.

Therefore the calculation of initial tension is carried out on the bridge with its sections without joining and entering the weight of this last deck section.

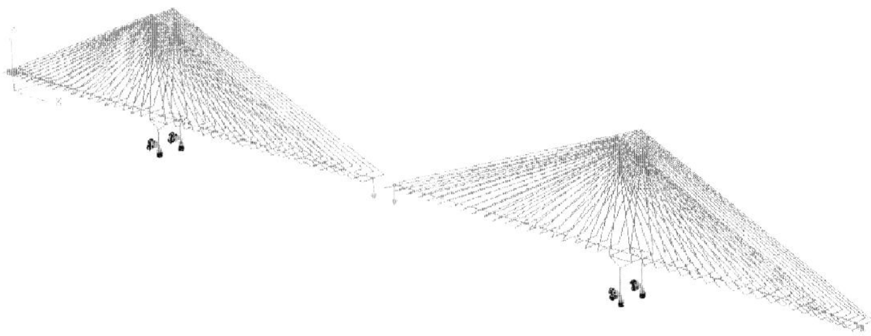

Figure 6: Structural model to obtain the initial tension forces.

These stresses are calculated as follows:

1. Deck displacement is calculated where cables are attached to the deck, which is produced by self-weight without initial stresses in the cables.
2. A unit force is applied to the first pair of cables and displacement is obtained at each point of the deck. This operation is repeated for the rest of the cable pairs.
3. Initial tension forces are obtained by resolving the following system of linear equations:

$$w_j^{pp} + \sum_{i=1}^{n} P_i \cdot w_j^{\,i} = 0 \qquad j = 1,\ldots,n$$

w_j^{pp} : deck displacement "j" due to the self-weight load

P_i : initial tension force in the cable pair, "i"

$w_j^{\,i}$: deck displacement at position, "j" due to the tension force of the cable pair, "i"

4.3 Optimization

The optimization process is carried out using an optimization module implemented in MATLAB. The function to perform this process is called "fmincon" that minimizes functions with various variables subject to inequality constraints.

This optimization subroutine requires certain input values such as design variables, constraints, and an objective function. For obtaining better results, it also provides gradients of the objective function and the design constraints. Those gradients are calculated by finite difference, the most costly part of the program.

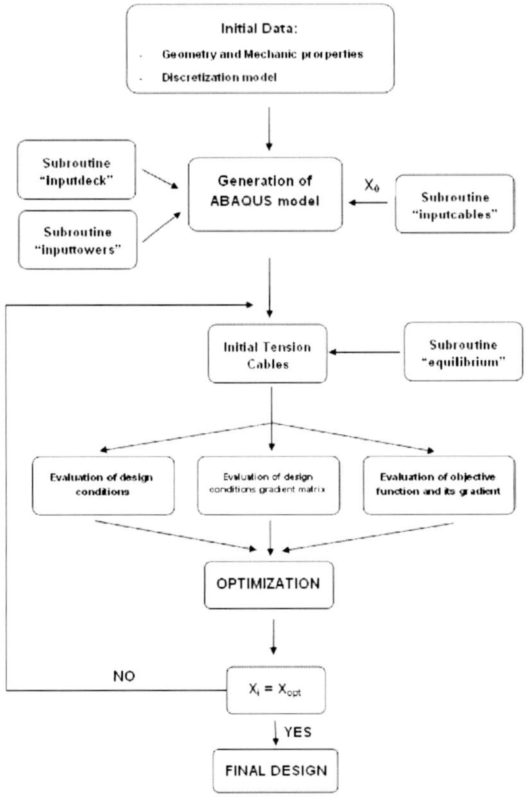

Figure 7: Flow chart of optimization process.

5 Application examples

The first example we present is a fictitious case of a bridge with the same longitudinal profile as the initial bridge described in section 2, but with noticeably reduced number of design variables (n=8).

The second example shows the solution obtained for the bridge configuration described at the beginning of the document. The number of variables in this case is much greater than the previous example (n=80).

In both examples the optimization of the cable is performed for three simple load cases as:

- Overload of use, 4 KN/m^2 on the left lateral span
- Overload of use, 4 KN/m^2 on the main span
- Overload of use, 4 KN/m^2 on the right lateral span

Obviously those load cases are not determinant for the limit state defined in the regulation, however, they serve to verify the program performance. The introduction of new load cases in the program does not present any difficulty since these cases simply need to be entered in the subroutine, "casos_carga".

5.1 Bridge with 16 cables

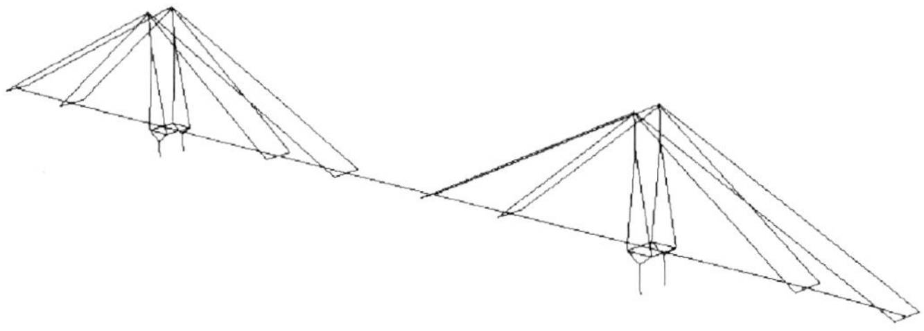

Figure 8: Structural model (16 cables).

- ***Design variables***

The design variables are the inverse of the areas of eight cables in the model. We use 0.1m^2 for the initial value of the area.

$$x_i^{\,0} = \frac{1}{A^0\_cable(i)} = \frac{1}{0.1} = 10 \quad i = 1,\dots,8$$

- ***Design constraints***

The design constraints considered for this example are summarized in the following equations:

$$\sigma^k(x_i) \le 800000 \,\frac{KN}{m^2} \quad i = 1,\dots,8 \quad k=1,2,3$$

$$\left|w^k(x_i)\right| \le \frac{L_1}{550} \quad i = 2 \quad k=1,2,3; \quad \left|w^k(x_i)\right| \le \frac{L_2}{550} \quad i = 3,\dots,6 \quad k=1,2,3$$

$$\left|w^k(x_i)\right| \le \frac{L_3}{550} \quad i = 7 \quad k=1,2,3; \quad \left|u^k{}_{Tower_j}\right| \le \frac{H_j}{550} \quad j=1,2 \quad k=1,2,3$$

where L1, L2 and L3 are the length of the left span, main span, and right span respectively and H1 and H2, the tower height. The dimensions are all in meter.

- ***Objective function***

The objective function is the total volume of steel cables that we intend to minimize.

$$F = 2\cdot\sum_{i=1}^{8} L_i \cdot \frac{1}{x_i}$$

The optimization process converges after 28 iterations giving results shown in figure 10 to 12.

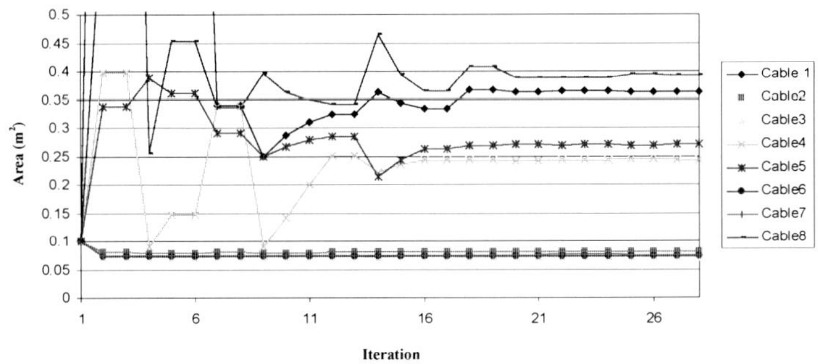

Figure 9: Evolution of the cable areas.

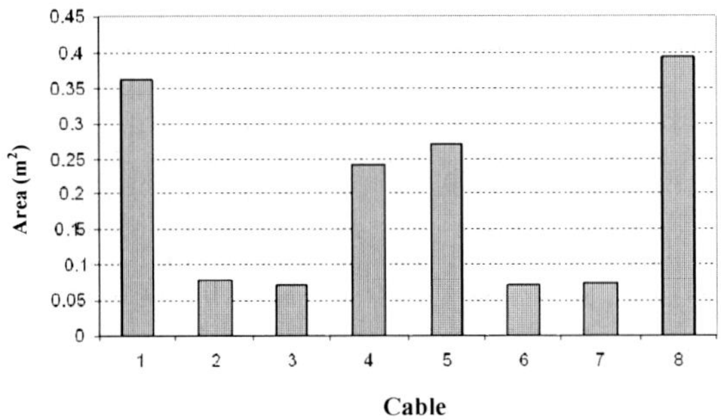

Cable	Initial Area	Optimal area
1	0.1	0.362
2	0.1	0.080
3	0.1	0.072
4	0.1	0.243
5	0.1	0.271
6	0.1	0.073
7	0.1	0.074
8	0.1	0.392

Figure 10: Distribution of cable area.

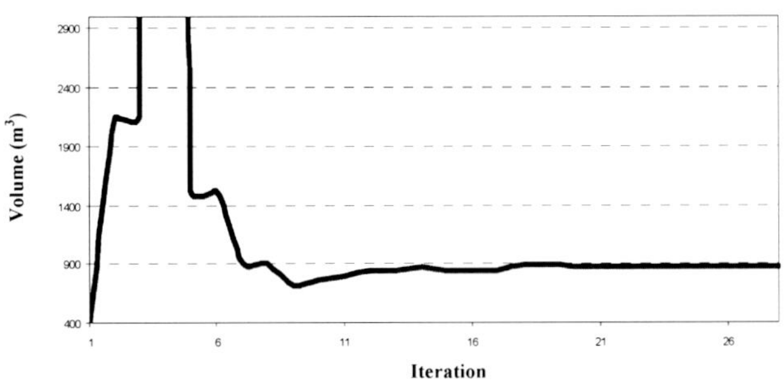

Figure 11: Evolution of the objective function.

The optimization results lead to larger area for cables attached to both ends of the deck just as occurs in the design of cable stayed bridges. Those cables are the ones that brace the towers. Likewise, the cables close to the center of the main span have a cross section larger than the rest, which can be observed in dimensioning of those bridges due to limiting the deck deflection.

In regard to the objective function, we can observe that the initial model has insufficient volume moving from 400m^3 to a value close to 900 m^3. Obviously these results do not correspond to any real cases since bridges are never designed with such small number of cables. We simply want to see the program performance.

After the optimization, we checked to see if all the design constraints are satisfied. The displacement constraint in the main span is active for the overload case on this section as well as the stress in the cables adjacent to those with the largest section. None of the constraints are violated.

5.2 Bridge with 160 cables

This example corresponds to the structural model described at the beginning of the document.

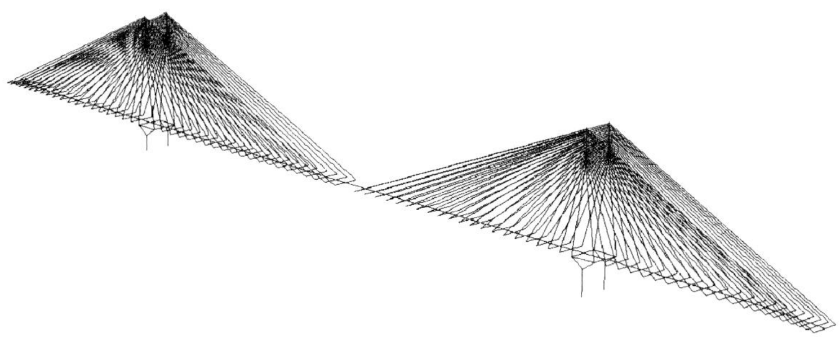

Figure 12: Structural model (160 cables).

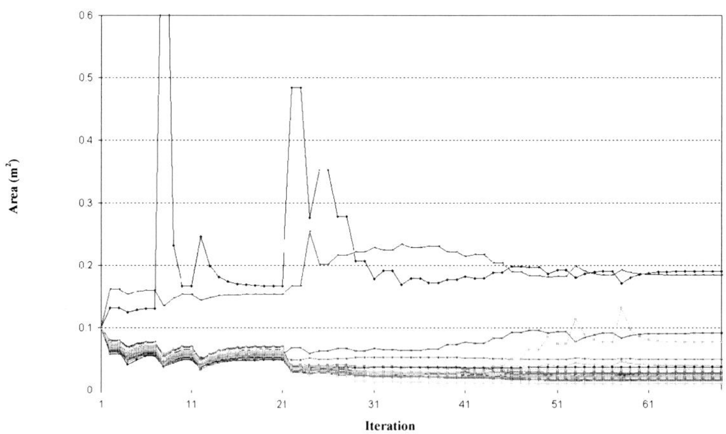

Figure 13: Evolution of the cross sections.

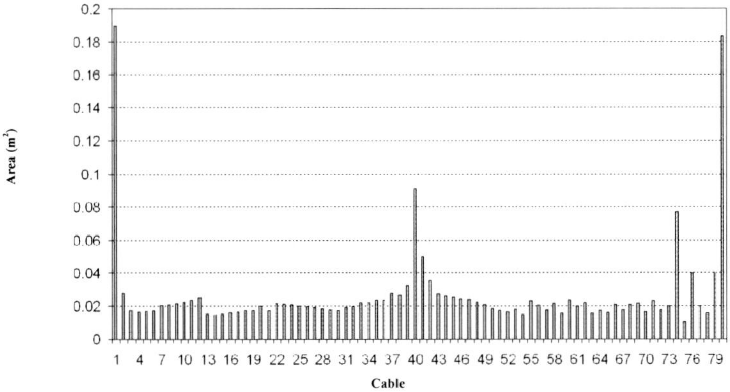

Figure 14: Distribution of areas.

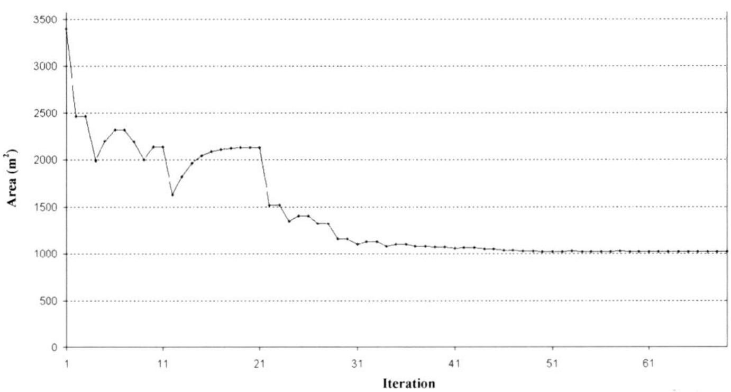

Figure 15: Evolution of the objective function.

Its geometrical and mechanical characteristics are identical to that of example 1 and it differs only for the number of cables and their locations.

Figure 13 shows the evolution of the cable areas during 69 iterations starting from an initial area of 0.1 m^2. It is different from the other model in terms that the cables at the extreme ends of the bridge have areas much larger than the rest, and two cables close to the center of the main span have approximately half of that value.

Figure 14 shows the area of each 80 cables after the optimization to give an idea of area distribution along the bridge.

Since the constraints are not violated in any moment, the design requirements are satisfied. The most limiting constraint is the displacement at the center of the main span for the distributed load on that span. Although the stress constraints are not active, some cables have the value close to the imposed limit.

Obviously we cannot assure that this optimum solution is unique, and there can be other local minimums. Nevertheless, the optimization proves to be useful to give more efficient design than that obtained by heuristic rules.

6 Conclusions

In this work, we have presented the results obtained from a study on the optimization of cable cross section of a cable stayed bridge.

Some conclusions can be drawn from the results:

1. Due to high degrees of indeterminacy of this type of structure, the optimization process is not monotonous presenting discontinuities in the design variables as well as in the objective function. A small modification in the design variables can affect the fulfillment of constraints far away from the location where the changes are produced.
2. We can check the results obtained in the optimization a posteriori by a finite element model. As shown in the examples, we can make sure that the results obtained satisfy the imposed requirements.
3. The program provides an idea to a design engineer how much cross sectional area is necessary for each cable under certain stress and displacement constraints. Since these values are not intuitive a priori at all, they serve as a design tool.
4. The distribution of optimized cable cross section along the bridge agrees with that of actual cable stayed bridges [6].
5. After the optimization process, it can be observed that the areas of cables attached to the extreme ends of the deck are much larger than the rest of the cables. Those cables serve to improve the longitudinal bracing of the towers and at the same time they have an important influence for vertical deck displacements.
6. The cross section of the cables located close to the center of the main span have larger area than the rest other than those at the extreme ends, due to fulfilling the deck displacement constraints.
7. The optimization process provides a result where the objective function has a minimum (local minimum), however, it does not guarantee that it is the best solution among all the possible ones (global maximum). Nevertheless as long as we achieve to reduce the steel volume in comparison to other conventional techniques, the optimization proves to be very useful to reduce material cost.
8. The developed program permits changes on the finite element model (geometry, mechanical properties, boundary condition, etc) in a simple way. Besides it permits to include load hypotheses within the optimization process whenever one is willing to pay the much higher computational cost. This cost is mainly due to obtaining the gradients of design constraints.

The conclusions drawn here on dimensioning of the cables comply exclusively with the terms in which the optimization process is defined. This study can be developed amply from many points of view such as:

1. Including geometry optimization to obtain optimum cable positioning on the deck
2. Including other structural elements such as the deck, towers, etc as design variables.
3. Considering other types of analysis in optimization such as buckling, vibration, dynamic analysis, etc.
4. Possibility of employing other types of optimization algorithms to achieve faster convergence

References

[1] N.J. GIMSING. *Cable Supported Bridges*. J. Wiley and Sons, Second Edition 1997.
[2] ABAQUS 6.8 Documentation.
[3] HERNÁNDEZ, S., *Métodos de Diseño Óptimo de Estructuras*.
[4] JASBIR S. ARORA, *Introduction to Optimum Design*, Second Edition, 2004.
[5] MATLAB Documentation.
[6] Honshu-Shikoku Bridge Authority, Japan. *The Tatara Bridge. Design and Construction Technology for the World's Longest Cable-Stayed Bridge. 1999.*

Heuristic optimization of prestressed concrete precast pedestrian bridges

J. V. Martí, F. González-Vidosa & J. Alcalá
School of Civil Engineering, Universidad Politécnica Valencia, Spain

Abstract

This paper deals with the economic optimization of prestressed concrete precast pedestrian bridges typically used in public works construction. These bridges are made of a precast concrete beam that integrates an upper reinforced concrete slab for the pedestrian traffic. The beam has a U-shape cross-section. Typical span lengths range from 20 to 40 m and the width ranges from 3.00 to 6.00 m. The study shows the efficiency of heuristic optimization by the simulated annealing (SA) and the threshold accepting (TA) algorithms. The evaluation of solutions follows the Spanish Code for structural concrete. Stress resultants and envelopes of these structures are computed by direct calculation. Design loads are in accordance with the national IAP Code for road bridges. The algorithms are applied to a typical pedestrian bridge of 40 m of span length and 6.00 m of width. This example has 59 discrete design variables for the geometry of the beam and the slab, materials in the two elements and active and passive reinforcement. The evaluation module includes the limit states that are commonly checked in design: flexure, shear, deflections, etc. The application of the SA and TA algorithms requires the calibration of the initial temperature and threshold, the number of variables modified in each iteration, the length of the Markov chains and the reducing coefficient. Each heuristic is run nine times so as to obtain statistical information about the minimum, average and deviation of the results. The best result has a cost of 27,586 euros for the SA algorithm and 27,570 euros for the TA algorithm. Finally, solutions and run times indicate that heuristic optimization is a forthcoming option for the design of real prestressed structures.
Keywords: economic optimization, heuristics, concrete structures, structural design.

WIT Transactions on The Built Environment, Vol 106, © 2009 WIT Press
www.witpress.com, ISSN 1743-3509 (on-line)
doi:10.2495/OP090111

1 Introduction

Artificial intelligence has dealt, since its appearance in the mid-fifties, with a variety of fields, such as automatic programming, the solution of constrained problems, operational research, planning, neuronal networks, etc. The design of structures is a problem of selection of design variables subject to structural constraints for which artificial intelligence is very much suited. Design variables include material grades, cross section dimensions and reinforcement. In spite of the potential capabilities of artificial intelligence, the present design of economic concrete structures is much conditioned by the experience of structural engineers. The traditional design of concrete structures aims at producing safe structures that sustain the loads and deformations, as well as the durability conditions of their intended life [1]. Also traditionally, designs have usually been based on the prior experience of the structural engineer, which has derived in a number of rules of thumb about cross-section dimensions and material grades based on sanctioned common practice. Once the structure is defined, it follows the analysis of stress resultants and the computation of passive and active reinforcement that satisfy the limit states prescribed by concrete codes. Should the dimensions or material grades be insufficient, the structure is redefined on a trial and error basis. Such process leads to safe designs, but the economy of the concrete structures is, therefore, very much linked to the experience of the structural designer.

Optimization methods deal with the minimization of the objective function, which is usually the cost or the weight of the structure. In addition, multiobjective optimization algorithms offer a range of possibilities about the study of two or several conflicting objectives. As regards concrete structures, there have been few applications when compared to steel structures [2]. In general terms, the methods of structural optimization may be classified into two broad groups: exact methods and heuristic methods. The exact methods are the traditional approach. They are usually based on the calculation of optimal solutions following iterative techniques of linear programming [3,4]. The second main group are the heuristic methods, whose recent development is linked to the evolution of artificial intelligence procedures. This group includes a broad number of search algorithms [5–8], such as genetic algorithms, simulated annealing, threshold accepting, tabu search, ant colonies, etc. These methods have been successful in areas different to structural engineering [9]. They consist of simple algorithms, but require a great computational effort, since they include a large number of iterations in which the objective function is evaluated and the structural constraints are checked. Among the first works of heuristic optimization applied to steel structures, the contributions of Jenkins and of Rajeev and Krishnamoorthy in 1992 are to be mentioned [10,11]. Both studies applied genetic algorithms to the optimization of the weight of steel structures. Regarding reinforced concrete (RC) structures, early applications in 1997 include the work of Coello et al [12], who applied genetic algorithms to the economic optimization of RC beams. Recently, there have been a number of RC applications [13–20], which optimize RC beams and building frames by genetic algorithms. Also recently, our research group has applied simulated annealing and threshold acceptance to the optimization of walls, frame bridges, building frames, vaults and bridge piers [21–25].

The structures object of this work are prestressed concrete (PC) precast pedestrian bridges typically used in public works construction. These bridges are made of a precast concrete beam that integrates an upper reinforced concrete slab for the pedestrian traffic (Figure 1). The beam has a U-shape cross-section. Typical span lengths range from 20 to 40 m and the width ranges from 3.00 to 6.00 m. The method followed in this work has consisted first in the development of an evaluation computer module where dimensions, materials and steel reinforcement have been taken as variables. These modules compute the cost of a solution and check all the relevant limit states. Simulated annealing and threshold accepting are then used to search the solution space.

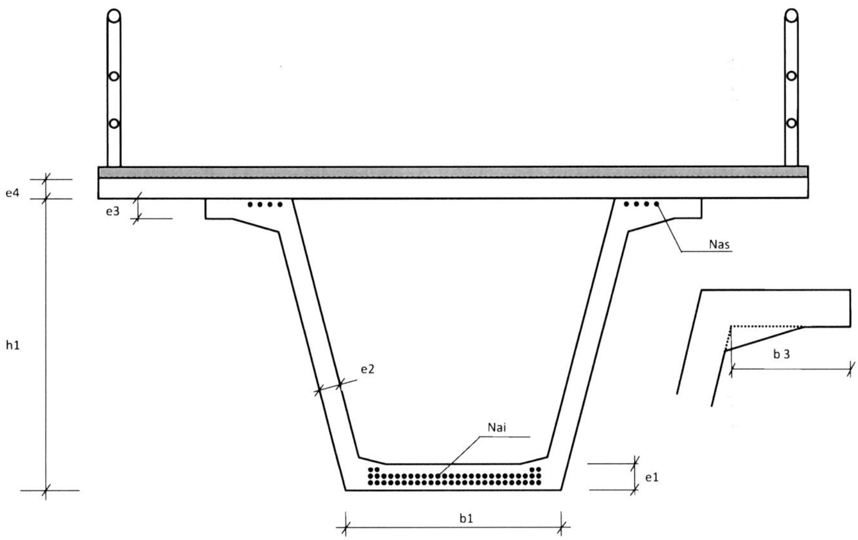

Figure 1: Geometric variables of the pedestrian PC precast bridge.

2 Annealing and threshold optimization procedures

The problem of structural concrete optimization that is put forward in the present work consists of an economic optimization. It deals with the minimization of the objective function F of expression (1), satisfying also the constraints of expressions (2).

$$F(x_1, x_2, \ldots x_n) = \sum_{i=1,r} p_i * m_i(x_1, x_2, \ldots, x_n) \qquad (1)$$

$$g_j(x_1, x_2, \ldots\ldots x_n) \leq 0 \qquad (2)$$

Note that the objective function in expression (1) is the sum of unit prizes multiplied by the measurements of construction units (concrete, steel, formwork, etc). And that the constraints in expression (2) are all the service and ultimate limit states that the structure has to satisfy. Unit prizes considered are given in Table 1.

Table 1: Basic prizes of the cost function of the reported structures.

Description	Unit Price in beam (€)	Unit Price in slab (€)
Kg of passive steel (B-500-S)	2.63	1.40
Kg of prestressing steel (Y-1860-S7)	3.38	NA
m of formwork	75.11	NA
m^2 of lower formwork	NA	30.00
m^3 of concrete HA-25	NA	64.99
m^3 of concrete HA-30	NA	69.95
m^3 of concrete HA-35	NA	74.03
m^3 of concrete HA-40	NA	79.12
m^3 of concrete HP-35	122.25	NA
m^3 of concrete HP-40	133.40	NA
m^3 of concrete HP-45	142.15	NA
m^3 of concrete HP-50	152.89	NA

The first search method used in this work is the simulated annealing (SA henceforth), which was originally proposed by Kirkpatrick et al. [26] for the design of electronic circuits. The SA algorithm is based on the analogy of crystal formation from masses melted at high temperature and let cool slowly. At high temperatures, configurations of greater energy than previous ones may randomly form, but as the mass cools, the probability of higher energy configurations forming decreases. The algorithm starts with a feasible solution randomly generated and a high initial temperature. The initial working solution is changed by a small random move of the values of the variables. The new current solution is evaluated in terms of cost. Greater cost solutions are accepted when a 0 to 1 random number is smaller than the expression $exp(-\Delta E/T)$, where ΔE is the cost increment and T is the current temperature. The current solution is then checked against structural constraints and if it is feasible, it is adopted as the new working solution. The initial temperature is decreased geometrically ($T=kT$) by means of a coefficient of cooling k. A number of iterations called Markov chains are allowed at each step of temperature. The algorithm stops when the temperature is a small percentage of the initial temperature or when there are no improvements in a number of Markov chains (typically 1% or 1-2 Markov chains). The SA method is capable of surpassing local optima at high-medium temperatures and gradually converges as the temperature reduces to zero. The SA method requires calibration of the initial temperature, the length of the Markov chains and the cooling coefficient. Adopted values for this work will be given below. The initial temperature was adjusted following the method proposed by Medina [27], which consists in choosing an initial value and checking whether the percentage of acceptances of higher energy solutions is typically 20-40 percent. If the percentage is greater than 40%, the initial temperature is halved; and if it is smaller than 20%, the initial temperature is doubled. Computer runs were performed a number of times so as to obtain minimum, mean and other statistical data of the results.

The second search method used in this work is the threshold accepting (TA henceforth), that was originally proposed by Dueck and Scheuer [28]. The algorithm starts with an initial randomly generated working solution, P_o, and an initial high threshold value for accepting solutions. TA then changes the solution by a move. A move is a small random variation up or down to the values of some of the variables that define the current solution. Given a current solution, a move is applied and, hence, a new solution is obtained. The new current solution, P_1, is accepted if it has a lower cost or when the cost increment is smaller than the current threshold value. The current solution is checked against structural constraints and if it is feasible, it is adopted as the new working solution P_o. The initial threshold is reduced geometrically and gradually by means of a reducing coefficient. A number of iterations at each threshold step are performed in the same way as for Markov chains of the simulated annealing method. TA method is capable of surpassing local optima and gradually converges as the threshold value reduces to zero. The method requires calibration of the initial threshold, of the length of the cycle chains and of the threshold-reducing coefficient. Adopted values for the two examples of this work will be given below. The initial threshold value was adjusted following the method proposed by Medina [27], which consists in choosing an initial value and checking whether the percentage of acceptances of higher cost solutions is typically between for instance 20-40 percent. If the percentage is greater than 40%, the initial threshold is halved; and if it is smaller than 20%, the initial threshold is doubled. Computer runs were performed 9 times so as to obtain minimum, mean and standard deviation of the random results.

3 PC precast pedestrian bridge example

The example studied relates to PC precast pedestrian bridges used in road construction [29]. The analysis includes 59 design variables. Fig. 1 shows the main geometrical variables considered in this analysis. Geometrical variables include 7 variables: the depth of the beam (h_1), the thickness of the slab (e_4), the width of the soffit of the beam (b_1) and the thickness of the bottom flange (e_1), the width and thickness of the top flanges of the beam (b_3 and e_3) and the thickness of the webs (e_2). Regarding material strength, 2 variables define the type of concrete of the slab and the beam. Prestressing is defined by 4 variables: the number of strands in the top flanges and the number of strands in the first, second and third layers of the bottom flange. Finally, 46 variables define a standard reinforcement set up for the beam and the top slab (Figure 2). All variables are discrete in this analysis. The total number of parameters is 15, the most important of which are the free span, the width of the top slab, the inclination of the webs, the dead and live loads acting on the bridge, the type of active and passive reinforcing steel, the ambient conditions and the partial coefficients of safety. Structural constraints considered followed standard provisions for Spanish design of this type of structure [30,31], that include checks of the service and ultimate limit states of flexure and shear for the stress envelopes due to the self weight and the traffic loads. Traffic load considered is a uniform distributed load of 4 kN/m^2. The dead load considered is a uniform distributed load of 2 kN/m^2. Stress resultants and reactions were calculated by a stiffness matrix program using a 2-D mesh with 20 bars and 21 sections. Deflections were limited to 1/1000 of the free span for the quasi-permanent combination. Fatigue of concrete and steel was not considered since this ultimate limit state is rarely checked in pedestrian bridges.

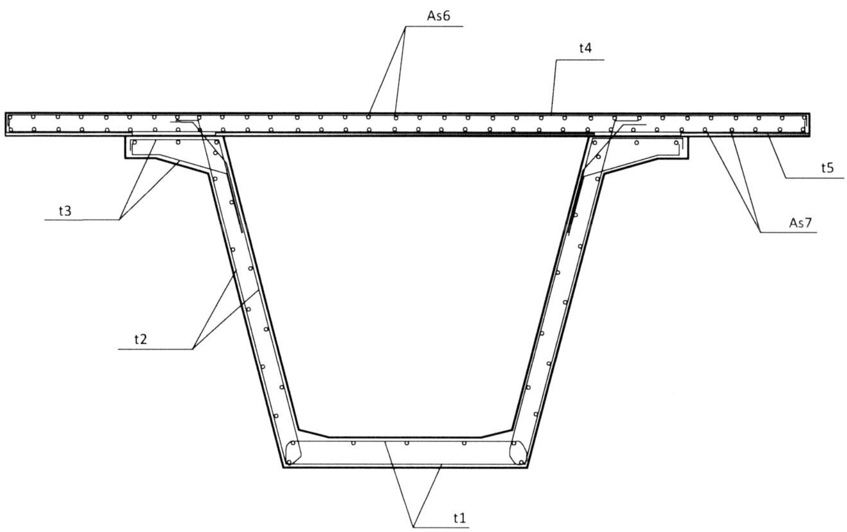

Figure 2: Passive reinforcement variables of the pedestrian bridge.

The SA-TA algorithms were programmed in Fortran 95 and applied to a pedestrian bridge of 40 m of span and a 6.00 m width. Typical SA-TA runs were about 25-15 minutes in an Intel Pentium IV processor of 3.00 GHz. The most efficient move found consisted of random variation of up to 10% of the 59 variables of the problem. The calibration of the simulated annealing parameters recommended Markov chains of 1000 iterations, a reducing coefficient of 0.95, an initial range of 40-60% of acceptances for the initial temperature and a stop criterion of 2% of the initial temperature and 2 chains without improvement. Figure 3 shows a typical evolution of the cost with the computing time for the SA algorithm. The minimum result has a cost of 27586.21 euros and the mean of the results is 29070.49, the deviation with respect to the best being 5.38%.

The calibration of threshold accepting parameters recommended chains of 500 iterations, a reducing coefficient of 0.95, an initial range of 70-90% of acceptances for the initial threshold and a stop criterion of 2% of the initial threshold and 2 chains without improvement. Figure 4 shows a typical evolution of the cost with the computing time for the TA algorithm. The minimum result has a cost of 27570.17 euros and the mean of the results is 30378.42, the deviation with respect to the best being 10.19%.

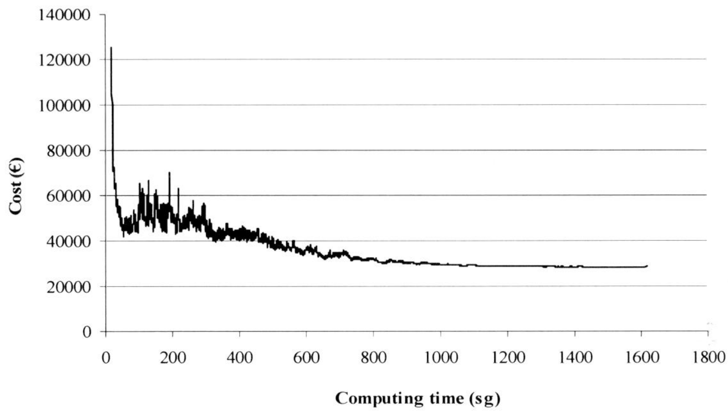

Figure 3: Typical cost evolution versus computing time for the SA algorithm.

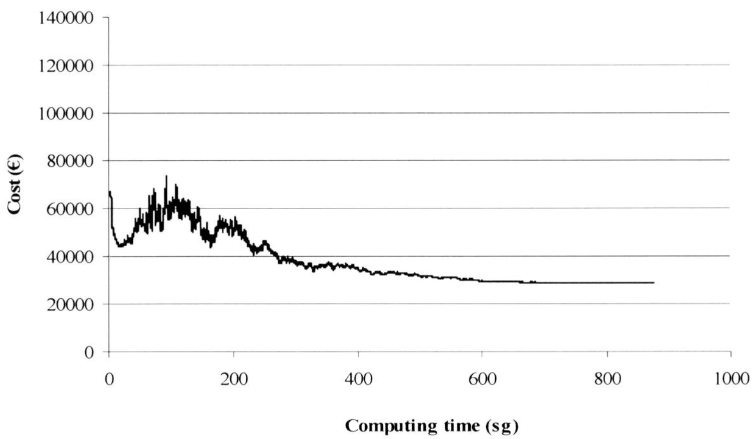

Figure 4: Typical cost evolution versus computing time for the TA algorithm.

It is worth noting that the SA algorithm outperforms the TA algorithm since it improves the deviation with respect to the best in the series of nine results.

Table 2 details the main results of the SA analysis for a pedestrian bridge of 40 m of span and a 6.00 m width.

Table 2: Summary of best pedestrian bridge.

Geometric variables	
h1	2.950 m
b1	1.180 m
b3	0.240 m
e1	0.150 m
e2	0.100 m
e3	0.180 m
e4	0.125 m
Nas	6ø0.6’
Nai	29ø0.6’
Concrete grades	
Beam	HA-35
Slab	HA-25
Reinforc. at support cross-section	
As6	24ø6
As7	24ø6
t1	12ø6/m
t2	12ø6/m
t3	12ø6/m
t4	8ø10/m
t5	16ø10/m

4 Conclusions

From the above work, the following conclusions may be derived:
- Two SA-TA algorithms have been implemented for the optimization analysis of PC precast pedestrian bridges.
- As regards the comparison of the SA-TA procedures, they have proved efficient search algorithms for the pedestrian bridge studied, although the SA appears more efficient regarding deviation with respect to the best.
- The example presented shows the potential applicability of local search heuristic procedures for the analysis of real PC structures.

References

[1] British Standards, BS 8110-1: 1997 Structural use of concrete. Part 1: Code of practice for design and construction, British Standards, London, 1997.

[2] Cohn M.Z. and Dinovitzer A.S., Application of structural optimization. ASCE Journal of Structural Engineering, **120**(2), pp 617–649, 1994.

[3] Hernández S. and Fontan A., Practical Applications of Design Optimization, WIT Press: Southampton, 2002.

[4] Fletcher R., Practical Methods of Optimization, Wiley: Chichester, 2001.

[5] Jones M.T., Artificial Intelligence Application Programming, Charles River Media: Hingham (Massachusetts), 2003.
[6] Holland J.H., Adaptation in natural and artificial systems, University of Michigan Press: Ann Arbor, 1975.
[7] Goldberg D.E., Genetic algorithms in search, optimization and machine learning, Addison-Wesley, 1989.
[8] Glover F. and Laguna M., Tabu Search, Kluwer Academic Publishers: Boston, 1997.
[9] Yepes V. and Medina J.R. Economic heuristic optimization for the heterogeneous fleet VRPHESTW. ASCE Journal of Transportation Engineering, **132**(4), pp 303–311, 2006.
[10] Jenkins W.M., Plane frame optimum design environment based on genetic algorithm, ASCE Journal of Structural Engineering, **118**(11), pp. 3103–3112, 1992.
[11] Rajeev S. and Krisnamoorthy C.S., Discrete optimization of structures using genetic algorithms, ASCE Journal of Structural Engineering, 118(5), pp. 1233–1250, 1992
[12] Coello C.A., Christiansen A.D. and Santos F., A simple genetic algorithm for the design of reinforced concrete beams, Engineering with Computers, **13**, pp. 185–196, 1997.
[13] Rajeev S. and Krisnamoorthy C.S., Genetic algorithm-based methodology for design optimization of reinforced concrete frames, Computer-Aided Civil and Infrastructure Engineering, **13**, pp 63–74, 1998.
[14] Koumousis V.K. and Arsenis S.J., Genetic algorithms in optimal design of reinforced concrete members, Computer-Aided Civil and Infrastructure Engineering, **13**, pp 43–52, 1998.
[15] Hrstka O., Kucerova A., Leps M. and Zeman J., A competitive comparison of different types of evolutionary algorithms, Computers and Structures, **81**, pp 1979–1990, 2003.
[16] Leps M. and Sejnoha M., New approach to optimization of reinforced concrete beams, Computers and Structures, **81,** pp 1957–1966, 2003.
[17] Lee C. and Ahn J., Flexural design reinforced concrete frames by genetic algorithm, ASCE Journal of Structural Engineering, **129**(6), pp 762–774, 2003.
[18] Camp C.V., Pezeshk S. and Hansson H., Flexural design reinforced concrete frames using a genetic algorithm, ASCE Journal of Structural Engineering, **129**(1), pp 105–115, 2003.
[19] Sahab M.G., Ashour A.F. y Toporov V.V., Cost optimisation of reinforced concrete flat slab buildings, Engineering Structures, **27**, pp 313–322, 2005.
[20] Govindaraj V. and Ramasamy J.V., Optimum detailed design of reinforced concrete continuous beams using genetic algorithms, Computers and Structures, **84**, pp 34–48, 2005.
[21] Gonzalez-Vidosa F., Yepes V., Alcala J., Carrera M. and Perea C., Simulated annealing optimization of walls, portal and box reinforced concrete road structures, In Proceedings of the Ninth International Conference on Computer

Aided Optimum Design in Engineering, Skiathos (Greece), pp 175–186, 2005.

[22] Carbonell A, Martínez FJ, Yepes V, Hospitaler A and González-Vidosa F., Threshold accepting optimization of road vaults and rectangular hollow bridge piers, In Proceedings of the Tenth International Conference on Computer Aided Optimum Design in Engineering (OPTI-2007), Myrtle Beach (South Carolina, USA), pp 141–150, 2007.

[23] Yepes V, Alcalá J, Perea C and González-Vidosa F, A parametric study of optimum earth retaining walls by simulated annealing, Engineering Structures, **30**(3), pp. 821–830, 2008.

[24] Perea C, Alcalá J, Yepes V, González-Vidosa F and Hospitaler A, Design of reinforced concrete bridge frames by heuristic optimization, Advances in Engineering Software, **39**(8), pp. 676–688, 2008.

[25] Payá I, Yepes V, González-Vidosa F and Hospitaler A, Multiobjective optimization of reinforced concrete building frames by simulated annealing, Computer-Aided Civil and Infrastructure Engineering, **23**(8), pp. 575–589, 2008.

[26] Kirkpatrick S., Gelatt C.D. and Vecchi M.P., Optimization by simulated annealing, Science, **220**(4598), pp. 671–680, 1983.

[27] Medina J.R., Estimation of incident and reflected waves using simulated annealing, ASCE Journal of Waterway, Port, Coastal and Ocean Engineering, **127**(4), pp. 213–221, 2001.

[28] Dueck G. and Scheuer T. Threshold accepting: A general purpose optimization algorithm superior to simulated annealing. Journal of Computation Physics, **90**, pp 161–175, 1990.

[29] Marti JV, Heuristic optimization of prestressed concrete precast pedestrian bridges (in Spanish), Research Report CST/GPRC-14, Dep. Construction Engg., Un. Politécnica Valencia, January 2009.

[30] M. Fomento, IAP-98. Code about the actions to be considered for the design of road bridges (in Spanish), M. Fomento: Madrid, 1998.

[31] M. Fomento, EHE. Code of Structural Concrete (in Spanish), M. Fomento, Madrid, 1998.

Reliability based design optimization of semi-rigid steel frames

M. A. Hadianfard
Department of Civil Engineering, Shiraz University of Technology, Iran

Abstract

In the conventional methods of structural optimization, the designers reduce the structural cost or weight without taking into account uncertainties in the materials, geometry and loading. In this way, the optimum solution may represent a lower level of reliability and thus a higher risk of failure. However, the objective of reliability based design optimization (RBDO) is to design structures that are economic and reliable. In this paper, external loads and member resistance are assumed to be independent random variables. To identify critical failure modes, the β-unzipping method is used and the probability of failure of each mode is calculated using the method of reliability analysis of parallel systems. By combination of the different failure modes as a series system, the probability of failure of the total structural system is estimated. The effects of the semi-rigid behavior of the beam-column connections are also considered in the RBDO problem. The RBDO problem is formulated by the expected costs or structural weight as the objective function and the reliability of the system as constraints. This constrained nonlinear optimization problem is changed in to an unconstrained optimization problem using the method of interior penalty function. Then, this new problem is solved by an unconstrained optimization technique. Some numerical examples are presented to demonstrate the applicability of the proposed methodology. These examples illustrate the importance of the effect of the semi-rigid behavior of the connections in the calculation of the minimum weight of the steel frames.

Keywords: reliability, structural optimization, steel frames, semi-rigid connections, probability of failure.

WIT Transactions on The Built Environment, Vol 106, © 2009 WIT Press
www.witpress.com, ISSN 1743-3509 (on-line)
doi:10.2495/OP090121

1 Introduction

The optimization of structures with respect to weight or cost is a well-known problem and it is the subject of many books, such as Kirsch [1] and Haftka and Gurdal [2]. However, the optimization of structures under reliability constraints is a complicated problem and it has been the subject of a few studies, such as Nakib and Frangopol [3]. The purpose of the reliability based design optimization (RBDO) is to produce reliable, and at the same token economical, structures. In the present study, by using a typical moment-rotation diagram, such as that described by Kishi et al. [4] and Hadianfard and Razani [5], the more realistic semi-rigid behavior of beam-column connections is considered in the structural analysis and in the RBDO. External loading, the resistance capacities of members and connection properties are assumed to be independent basic random variables, with normal distribution. The β-unzipping method, as described by Thoft-Christensen and Murotsu [6], is used to identify critical failure modes in the structural system. For each critical mode the probability of failure is calculated by using the method of reliability analysis of parallel systems. Then, by combination of the different failure modes as a series system and by using bounding techniques, the probability of failure of the overall structural system is estimated. In the optimization problem, the objective function is the weight of the structure, including the weight of the members and the weight of the connections, while the constraints are both deterministic and probabilistic.

2 Structural analysis of semi-rigid steel frames

The conventional analysis and design of steel frames are usually carried out under the assumption that the beam-column connections are either fully rigid or ideally pinned. However, the actual behavior of the beam-column connections in the steel frames is semi-rigid. The relationship between the beam end moment (*M*) and the relative change in angle (θ_r) can be described by means of a typical moment-rotation curve. Some popular $M-\theta_r$ relations, such as linear, polynomial, exponential, and power models, are explained by Chen and Lui [7].

In the present research, the power model of Kishi et al. [4] and Chen and Lui [7] is used for describing the $M-\theta_r$ the relation of the connection with angles. The $M-\theta_r$ relation in this model is shown in eqn (1).

$$\theta_r = \frac{M}{R_{ki}[1-(\frac{M}{M_u})^n]^{1/n}} \tag{1}$$

In which R_{ki} is the initial connection stiffness, M_u is the ultimate moment capacity of the connection, and n is a shape parameter of the $M-\theta_r$ curve. The two parameters, R_{ki} and M_u are evaluated analytically based on the size and geometry of the connection, while the shape parameter n is obtained by curve-fitting technique. The effects of the connection flexibility are modeled by attaching rotational springs with stiffness modulus R_i and R_j at the ends i and j of a beam. For the sake of simplicity and for reducing the time of analysis, it is assumed that the rotational spring are linear with the stiffness modulus equal to $0.5R_{ki}$ The stiffness matrix of a beam with rotational springs at the ends is function of the R_i and R_j as given by Hadianfard and Razani [5]. For columns, the stiffness matrix takes the usual form. The beams and the columns stiffness matrices can be assembled in the usual manner to form the stiffness matrix of the structure.

3 Identification of critical failure modes

The reliability of a structure is denoted by the factor $\Re$ which is defined by:

$$\Re = 1 - P_f \tag{2}$$

Where, P_f is the probability that the structure will fail during the specified reference period. For a fundamental case (a structure with only two independent random variables R and S, where R is resistance variable and S is load effect variable) the reliability index β is defined by eqn (3).

$$\beta = \frac{\mu_M}{\sigma_M} \tag{3}$$

where M=R-S is called the safety margin and μ_M, σ_M are the mean value and standard deviation of M. The probability of failure (P_f) for this fundamental case can be calculated from eqn (4).

$$P_f = \Phi(-\beta) \Leftrightarrow \beta = -\Phi^{-1}(P_f) \tag{4}$$

where Φ is standard normal distribution function.

The β-unzipping method is one of the methods by which the reliability of the structures with ductile or brittle elements can be estimated at a number of different levels. The systems reliability at mechanism level is defined as the reliability of a series system with the significant mechanisms as elements. For generating the Fundamental mechanisms, the structure must be modeled by n failure elements. At first it is assumed that, the failure is accrued in critical failure element l with the lowest reliability index, and structure is modified by removing the corresponding failure elements and adding a pair of so-called fictitious loads (F_l) (normal forces or moments for ductile elements). The modified structure is then reanalyzed and new reliability indices are calculated for the others failure elements. If the β for failure element m was the lowest β-values. Then the failure is assumed in the elements l and m and the structure is modified and reanalyzed. In this way new β-values are calculated for all failure elements except l and m and the smallest β-value is calculated (for example in failure element r). In this way, 3 failure elements l, m and r are selected as part of failure tree. The same way can be used for selecting other failure elements of the failure tree. Formation of a mechanism can be unveiled by the fact that the corresponding stiffness matrix is singular. By this procedure a number of mechanisms with different numbers of failure elements will be identified. In this way, several re-analyses of the structure are necessary.

4 Estimation of the probability of failure

The reliability of a structural system can be estimated on the basis of a modeling by a series system where the elements in the series system are parallel systems (failure modes). Probability of failure for each parallel system (P_{fp}) can be calculated by Ditlevsen's bounds. In this bounding method the lower and upper bounds are as following:

$$P_{fp(u)} = 1 - P[S_1] - \sum_{i=2}^{p} Max\{0, \sum_{j=1}^{i-1} P[\bar{S}_j \cap S_i] - (i-2)P[S_i]\} \tag{5}$$

$$P_{fp(l)} = 1 - P[S_1] - \sum_{i=2}^{p} \underset{j\in\{1,2,\dots,i-1\}}{Min}\{P[\bar{S}_j \cap S_i]\} \tag{6}$$

where, $P_{fp(u)}$ is upper bound and $P_{fp(l)}$ is lower bound of the probability of failure, M_i is safety margin for failure element *i*, $P[S_i]$ is probability of occurrence of non-failure state in the failure element *i*, $(P(M_i \succ 0))$ and $P[\bar{S}_i]$ is probability of occurrence of failure state in the failure element *i*, $(P(M_i \le 0))$ and *p* is total number of failure elements. In these equations the $P[\bar{S}_i \cap S_j]$ can be calculated as below:

$$P[\bar{S}_i \cap S_j] = P[(M_i \le 0) \cap (M_j > 0)] = 1 - P[S_i] - P[\bar{S}_i \cap \bar{S}_j] \tag{7}$$

where $P[\bar{S}_i \cap \bar{S}_j]$ is joint probability and it can be estimated as following:

$$Max(P_1, P_2) \le P_{ij} = P[\bar{S}_i \cap \bar{S}_j] \le P_1 + P_2 \tag{8}$$

$$P_1 = \Phi(-\beta_i)\Phi(-\gamma_j), \quad P_2 = \Phi(-\beta_j)\Phi(-\gamma_i) \tag{9}$$

$$\gamma_i = \frac{\beta_i - \rho_{ij}\beta_j}{\sqrt{1-\rho_{ij}^2}}, \quad \gamma_j = \frac{\beta_j - \rho_{ij}\beta_i}{\sqrt{1-\rho_{ij}^2}} \tag{10}$$

where ρ_{ij} is correlation coefficient as defined by eqn (11).

$$\rho_{ij} = \frac{Cov[M_i, M_j]}{\sigma_{Mi}\sigma_{Mj}} \qquad i \neq j \tag{11}$$

For calculation of probability of failure of the structural system (series system (P_{fs})) the simple bounds method (Cornell's bounds) as defined by eqn (12) can be used.

$$\underset{i=1}{\overset{n}{Max}}(P_i) \le P_{fs} \le 1 - \prod_{i=1}^{n}(1 - P_i) \tag{12}$$

where *Pi* is probability of failure of the element i, $(P(M_i \le 0))$.

5 Formulation of RBDO problems

The optimal design problem can be formulated mathematically as:

$$\begin{cases} Minimize & W(X_1, X_2, ..., X_n) \\ Such\ that & P_{fs}(X_1, X_2, ..., X_n) \le P_{fa} \end{cases} \tag{13}$$

Where, W is weight of the structural system consist of weight of the structural members (*Ws*) and weight of the connections (*Wc*). The parameter *Xi* is design variable, P_{fs} is probability of failure of the structural system and P_{fa} is a specified allowable probability of failure.

The structural weight can be obtained from eqn (14).

$$W = W_S + W_C = \sum_{i=1}^{n} \rho_i L_i A_i + \sum_{i=1}^{m} \rho_i V_{Ci} \tag{14}$$

where ρ_i, l_i, A_i are the density, length and the cross-sectional area of the member i, and V_{Ci} is the volume of the connection i, and *n*, *m* are the total number of members and connections respectively. In a structural system with constant density for all members and connections, the objective function can be shown as:

$$Minimize \quad V(A_i, K_i, M_{ui}) = \sum_{i=1}^{n} L_i A_i + \sum_{i=1}^{m} V_{Ci}(K_i, M_{ui}) \tag{15}$$

Then the design variables are: A_i, K_i, M_{ui} where, *Ki* and *Mui* are the beam-column connection properties (stiffness and moment capacity).

The volume of the connections (V_{Ci}) is function of the type, shape, flexibility, capacity and the others properties of the connections. The type and the shape of the connections must be selected before the optimization process, and the others properties can be obtained from the $M - \theta_r$ relation of the connection. For reducing the number of the design variables, the moment capacity of the connection can be considered as a function of connection stiffness (*Ki*). Then the design variables are only *Ai* and *Ki*, and these variables can be obtained from solving the RBDO problem.

6 Solving the optimization problem

The constrained nonlinear optimization problem of eqn (13) can be changed in to an unconstrained optimization problem using method of interior penalty function [1]. In the penalty function method, a penalty term that takes care of the constraints is added to the original objective function.

Consider the original optimization problem as eqn (16).

$$\begin{cases} Minimize & f(X) \\ Such\ that & g_j(X) \le 0 \quad j = 1,2,...,m \end{cases} \tag{16}$$

where *f(x)* is objective function and $g_j(X), (j = 1,2,...,m)$ are constraints of the optimization problem.

Thus, the transformed problem is as following:

$$Minimize\ \phi(X, r_k) = f(X) - r_k \sum_{j=1}^{m} \frac{1}{g_j(X)} \qquad k = 1,2,3,... \tag{17}$$

where r_k is the penalty parameter.

The solution process is as follows: Initial value of *{X}* and *r* are chosen. *{X}* must lie in the feasible region and *r* must be a positive number. For the given r_k, the objective function $\phi(X, r_k)$ is minimized to obtain *{X*}* and the convergence criterion of *{X*}* to the optimum is checked. If it is not satisfied, the parameter *r* is reduced ($r_{k+1} = cr_k,\ c < 1$) and starting from *{X*}* the function $\phi(X^*, r_{k+1})$ is minimized again. The step of reducing *r* and minimizing $\phi(X, r_k)$ are repeated until the convergence criterion is satisfied.

In each step of the penalty function technique, the obtained unconstrained optimization problem can be solved by using a direct search method as Powell or Rosenbrock method. For example the solution process for the Powell's method is as follows: 1-Choose an initial vector *{X}* and *n* initial independent directions *{Sq}, q=1,2,...,n.* 2- Select new vector *{Y}* equal to *{X}*. 3- Find α^* to minimize $F(\{X\}+\alpha\{Sq\})$ and set $\{X\} \leftarrow \{X\} + \alpha^* \{S_q\}$ for *q=1,2,...,n.* 4-Set $\{S_{n+1}\} \leftarrow \{X\} - \{Y\}$ and find α^* to minimize $F(\{X\} + \alpha\{S_{n+1}\})$, and set $\{X\} \leftarrow \{X\} + \alpha^* \{S_{n+1}\}$. 5- Replace $\{S_q\} \leftarrow \{S_{q+1}\}$ for *q=1,2,...,n.* 6- Repeat from step 2.

7 Numerical calculations and discussion

In this section some examples of numerical calculations are presented. These examples show applicability of the proposed methodology and importance of the consideration of semi-rigid behavior of the connections in the RBDO problems.

Example 1: A steel frame, as shown in figure 1, is assumed to be under vertical and lateral loads. External loads are random variables with mean values as given in figure 1. Also member resistances and moment capacities of the connections are random variables. All of the random variables are normally distributed with coefficient of variation equal to 0.1.

It is assumed that the expected value for yield strength of the steel is $\sigma_y = 240$ *Mpa.*

In this example, nine failure elements are considered in the reliability analysis. Four failure elements are in the beam, namely a yield hinge possibility at each end of the beam and another yield hinge at middle of the beam and the possibility of failure in axial force. Also each column has two failure elements, namely a yield hinge possibility at the end of the column (near the beam) and the possibility of failure in axial force. Another failure element is possibility of failure in axial force in the bracing element.

The RBDO problem can be formulated as following:

$$Minimize\quad V(A_i, K_i) = V_S(A_i) + V_C(K_i) = \sum_{i=1}^{n} L_i A_i + \sum_{j=1}^{m} V_{Cj}(K_j) \tag{18}$$

Such that:

1) $\frac{P_{fs}}{P_{fa}}-1\leq 0$, 2) $1-\frac{A_i}{A_{i,\min}}\leq 0$, $i=1,2,...,n$

3) $1-\frac{K_j}{K_{j,\min}}\leq 0$, $j=1,2,...,m$, 4) $\frac{K_j}{K_{j,\max}}-1\leq 0$, $j=1,2,...,m$

where A_i, L_i are area and length of the member i, K_j is rotational stiffness of the connection j ($K_j = 0.5R_{ki}$), $A_{i,min}$ is minimum cross sectional area of the member i (it is selected by designer before optimization). The parameters $K_{j,min}$, $K_{j,max}$ are minimum and maximum rotational stiffness of the connection j. These values can be selected with respect to the shape and type of the connections by using appropriate $M-\theta_r$ relation. By selecting the cross-sectional area and the shape of the section (for example: IPE, IPB, UNP and etc.) the others geometrical properties of the section, such as: moment of inertia, plastic modulus and etc. can be evaluated.

It is assumed that the connection weight (volume) is function of the connection stiffness. Then for fully pinned connection, the weight of the connection is minimum and for fully rigid connection, it is maximum. Relation between the connection volume (V_{cj}) and the connection stiffness (K_j) can be written as below:

$$V_{Cj}=\alpha_j[V_{j,\min}+C_j(K_j-K_{j,\min})]\leq V_{j,\max}\ ,\quad C_j=\frac{V_{j,\max}-V_{j,\min}}{K_{j,\max}-K_{j,\min}} \tag{19}$$

where $V_{j,\min}$, $V_{j,\max}$ are the minimum and maximum volume of each connection, and α is constructional factor $(\alpha\geq 1)$. This factor can be used for consideration the difficulties in the construction of the connections.

Also relation between the moment capacity (M_u) and the stiffness (K) of the connection can be written as:

$$M_u = M_{u,\min}+\frac{M_{u,\max}-M_{u,\min}}{K_{\max}-K_{\min}}(K-K_{\min}) \tag{20}$$

where $M_{u,\min}$, $M_{u,\max}$ are minimum and maximum moment capacities of the connections (it is selected by designer).

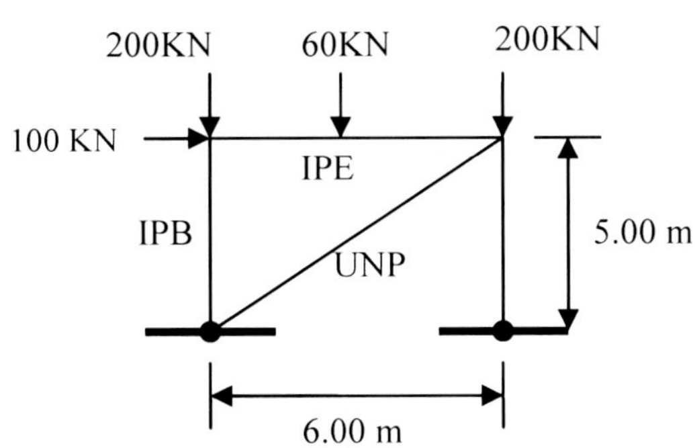

Figure 1: Steel frame of the example 1.

Some different types of the beam-column connections with angles are considered in this example and for each case the RBDO problem is solved.

a) Frame with fully pinned connections: In this ideal case, the initial stiffness and the moment capacity of the connections are zero and connection weight (volume) is at least. The following limitations are considered in this example.

$$A_{c,\min} = 26\ cm^2,\quad A_{b,\min} = 10\ cm^2,\quad A_{br,\min} = 12\ cm^2$$

$$K_{\min} = K_{\max} = 0,\quad M_{u,\min} = M_{u,\max} = 0$$

$$V_{c,\min} = 100\ cm^3,\quad V_{c,\max} = 200\ cm^3,\quad \alpha = 1.20,\quad P_{fa} = 0.001$$

where *Ac, Ab, Abr* are cross-sectional area of the columns, beam and bracing. Initial values for the design variables are selected as:

$A_c = 50\ cm^2,\quad A_b = 90\ cm^2,\quad A_{br} = 13\ cm^2$

The optimum solution can be found by solving the RDBO problem, and the results are as: $A_c^* = 26\ cm^2,\quad A_b^* = 52.2\ cm^2,\quad A_{br}^* = 12\ cm^2$.

In this case the minimum volume of the structure without consideration the connection volume is $V_{\min,1} = 66692\ cm^3$, and with consideration the volume of the connections is $V_{\min,2} = 67052\ cm^3$.

b) Frame with single web angle connections: In conventional ways this type of the connection was assumed to behave as a fully pinned connection, but the real behavior of the connection is semi-rigid and this behavior can be shown by the Kishi and Chen power model (eqn (1)).
For example in a web connection by angle $L100 \times 100 \times 10mm$ with length *l=12cm* the parameters of Kishi and Chen model are as:

$R_{ki} = 1.92 \times 10^3\ KN.m/rad,\quad M_u = 4.6\ KN.m$. In this case the volume of each connection is about $V_{ci} = 230\ cm^3$. The upper and lower bounds for these parameters can be obtained by changing the size and the length of the connection angle. Therefore the following limitations may be assumed for the connections.

$$K_{\min} = 0.25 \times 10^3,\ K_{\max} = 3.5 \times 10^3,\ M_{u,\min} = 2,\ M_{u,\max} = 14,\ V_{c,\min} = 150,\quad V_{c,\max} = 400,\quad \alpha = 1.20$$

By selecting the suitable initial values as: $A_c = 40,\ A_b = 60,\ A_{br} = 20,\ K = 1 \times 10^3$ the optimum solution can be found as:

$$A_c^* = 26\ cm^2,\ A_b^* = 48.6\ cm^2,\ A_{br}^* = 12\ cm^2,\ K^* = 1.31 \times 10^3,\ V_{\min,1} = 64532,\ V_{\min,2} = 65012$$

The optimum volume in this manner is less than the optimum volume of the case (a).

c) Frame with double web angles connections: In this type of the connection the limits on the connection properties are two times the values of the case (b). The optimum solution is given in table 1.

d) Frame with top and seat angles connections: The properties of these semi-rigid connections can be found by Kishi and Chen model. For example, in a connection by the angles $L100 \times 100 \times 10mm$ with length *l=12cm* and with beam depth equal to 24 cm, the initial stiffness is $R_{ki} = 9.79 \times 10^3\ KN.m/rad$ and the moment capacity is $M_u = 27.4KN.m$ and the volume of each connection is about $V_{ci} = 460cm^3$. Then the constraints on the connection properties may be selected as:

$K_{min} = 3.0\times10^3$, $K_{max} = 9.0\times10^3$, $M_{u,min} = 15$, $M_{u,max} = 45$, $V_{c,min} = 300$, $V_{c,max} = 800$

The optimum solution of this case is given in table 1.

Table 1: Results of the RBDO for the frame of example 1.

Connection type	A_c^* (cm^2)	A_b^* (cm^2)	A_{br}^* (cm^2)	$V_{min,1}$ (cm^3)	$V_{min,2}$ (cm^3)	K_{opt} $(KN.m/rad)$
a) fully pinned	26	52.5	12	66692	67052	0
b) single web angle	26	48.6	12	64532	65012	1000
c) double web angle	26	46.7	12	63429	64389	2000
d) seat angle	26	46.5	12	63273	64392	5000
e) seat angle + web angles	26	46.5	12	63273	65937	7000
f) fully rigid	26	46.5	12	63273	71790	∞

e) Frame with top and seat angles with double web angles connections: For a connection by angles $L100\times100\times10mm$ with length *l=12cm* and with beam depth equal to 24 cm, the initial stiffness is $R_{ki} = 14.6\times10^3\ KN.m/rad$ and the moment capacity is $M_u = 48.5KN.m$. Then the constraints on the connection properties may be selected as:

$K_{min} = 5.0\times10^3$, $K_{max} = 12\times10^3$, $M_{u,min} = 30$, $M_{u,max} = 70$, $V_{c,min} = 600$, $V_{c,max} = 1500$

The optimum solution of this case is given in table 1.

f) Frame with fully rigid connections: In this ideal case, it is assumed that the initial stiffness of the connections is infinite (a large number as $K = 1\times10^6\ KN.m/rad$) and the moment capacity of the connections is greater than the moment capacity of the connected beam. Also in this case the connection weight (volume) is maximum and it is about $V_{ci} = 1800cm^3$. Because of difficulty in the construction of the rigid connections, the parameter α is assumed to be $\alpha=1.30$. The results of the RBDO are shown in the table 1. It can be seen from the table 1 that by increasing the connection stiffness, the weight (volume) of the structural members will be decreased but the weight of the connections will be increased. Then at the optimum connection stiffness, the total weight of the structural system (consist weight of the connections) will be at least. The case c (connection by double web angles) is an optimum manner between all of the cases. Also this table appears that the weight of the overall structure in the cases of fully pinned and fully rigid is greater than the weight of the other cases. Therefore, ignoring the weight and the behavior of the semi-rigid connections can lead to the unrealistic optimum solution in the steel frames.

Example 2: A steel frame as shown in figure 2 is assumed to be under uniform vertical loads with expected value 15 KN/m. It is assumed that the coefficient of variation for all of the random variables is equal to 0.1. All beam-column connections are similar and made of angles. Basic design variables are rotational stiffness of the connections (K) and cross-sectional area of the members as: Ac1: corner columns in the first story, Ac2: middle column in the first story, Ac3: corner columns in the second story, Ac4: middle column in the second story, Ab1: beams in the first story, Ab2: beams in the second story.

The constraints of this example are: $A_{c,\min} = 26\ cm^2$, $A_{b,\min} = 10\ cm^2$, $P_{fa} = 0.001$ ·

The results of the RBDO for different type of the connections are shown in table2. This table indicates that the case (d) is an optimum case, and the weight (volume) of the overall structure in this case is at least.

Table 2: Results of the RBDO for the frame of example 2.

Connection type	$A^{\bullet}_{c1}$ (cm^2)	$A^{\bullet}_{c2}$ (cm^2)	$A^{\bullet}_{c3}$ (cm^2)	$A^{\bullet}_{c4}$ (cm^2)	$A^{\bullet}_{b1}$ (cm^2)	$A^{\bullet}_{b2}$ (cm^2)	$V_{\min,1}$ (cm^3)	$V_{\min,2}$ (cm^3)	K_{opt} $(KN.m/rad)$
(a)	26	26	26	26	31	31	110400	111840	0
(b)	26	26	26	26	27	27	103680	105600	1000
(c)	26	26	26	26	25.5	25.5	101460	105300	2000
(d)	26	26	26	26	21.5	22.5	94238	98712	5000
(e)	26	26	26	26	19	21	90485	99912	7000
(f)	26	26	26	26	16.5	18.5	86100	104820	∞

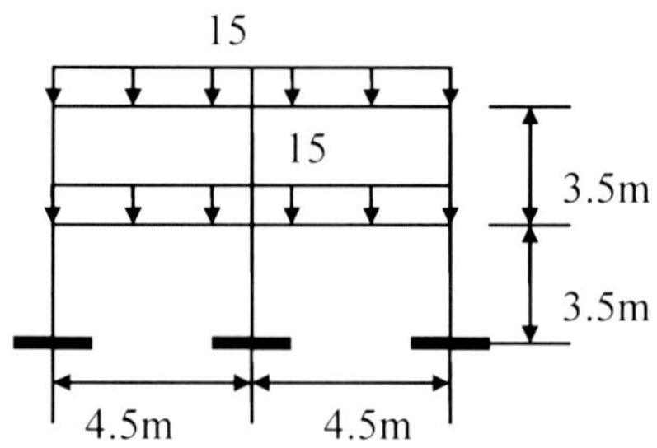

Figure 2: Steel frame of example 2.

Example 3: In this example a two-story steel frame is considered as shown in figure 3. The mean values of the loads variables are as: *P1=169 KN, P2=89 KN, P3=116 KN, P4 =31 KN*. The coefficient of variation for *P2, P3* and *P4* is equal to 0.25 and for the others loads and design variables is equal to 0.15. As shown in the figure 3 the number of potential yield hinge is n=19. In this example six design variables are considered. One is rotational stiffness of the connections (K) and the others are cross-sectional area of the members as: Ac1: columns in the first story, Ac2: columns in the second story, Ab1: left beam in the first story, Ab2: right beam in the first story, Ab3: beam in the second story.

All connections are end plates with moment capacity equal to the plastic moment of the connected beam. This type of the connection is usually assumed to be fully rigid. However, the actual behavior of this connection is semi-rigid and can be modeled by Fry and Morris polynomial model as described by Chen and Lui [7]. For the end plate connection with thickness tp=2.5 cm and height dg=50 cm, the initial stiffness of the connection is equal to $R_{ki} = 8.2 \times 10^4 \, KN.m/rad$. Two types of the connection behavior are considered. One is fully-rigid (ideal behavior) and another is semi-rigid behavior (actual behavior). Many different allowable probability of failure are considered. Results of the optimization process are presented in table 3 for some values of the P_{fa}. Also the variation of the optimum volume is presented with respect to the P_{fa} in figure 4. Results show that, the optimum weight (volume) of the overall structure in the case of fully-rigid is less than the case of semi-rigid. Then the assumption of fully rigid connections in the RBDO is unsafe and may lead to unrealistic results.

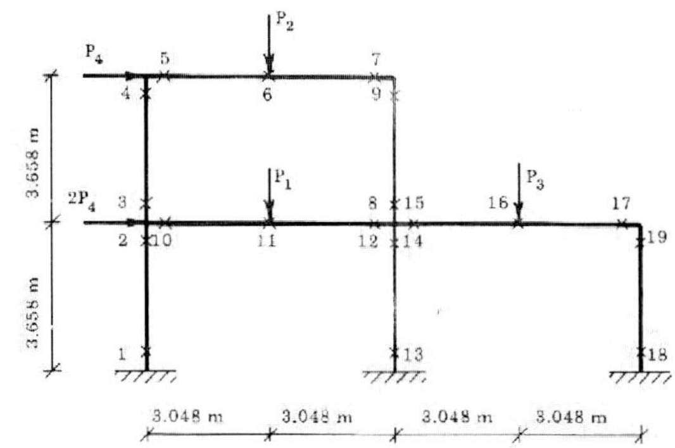

Figure 3: Steel frame of example 3.

Table 3: Results of the RBDO for the frame of example 3.

Connection Type	$P_{fa} = 0.001$						$P_{fa} = 0.0001$					
	C1	C2	B1	B2	B3	Vmin	C1	C2	B1	B2	B3	Vmin
Fully-rigid	601.1	64.4	70.6	74.8	72.7	267473	82.7	80.1	79.4	78.4	80.6	316135
Semi-rigid	70.1	55.3	81.6	76.9	73.1	280019	83.5	82.2	89.4	88.5	85.3	333667

8 Conclusions

In this research, the realistic semi-rigid behavior of the beam-column connections is considered in the RBDO problem. The total structural weight (consist of the connections weight) is minimized while the probability of failure of the structure did not exceed a prescribed acceptable value. The numerical examples indicate the applicability of the proposed method. Also theses examples indicate the importance of the assumption of semi-rigid behavior of the connections in the RBDO. The assumption of fully-pinned connections is usually safe and over design but the assumption of fully-rigid connections is usually unsafe and under design. Further by the proposed method; the optimum stiffness of the connections can be evaluated. This optimum connection stiffness can lead to minimum weight of the structure.

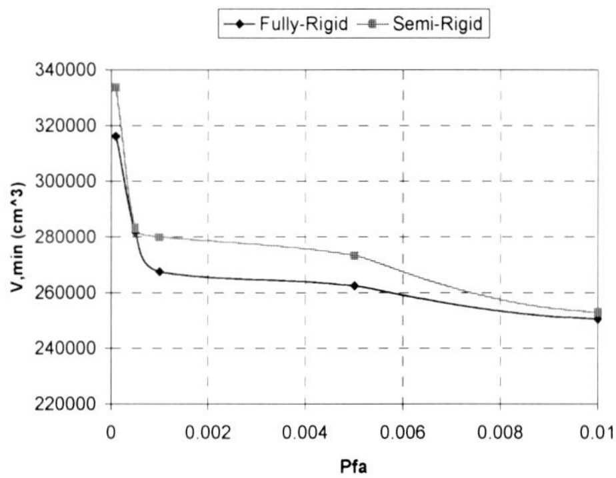

Figure 4: Optimum volume of the frame of example 3 for the different P_{fa}.

References

[1] Kirsch, U., *Optimum Structural Design*, McGraw-Hill Book Company, 1981.
[2] Haftka, R.T., Gurdal, Z., *Elements of Structural Optimization*, Kluwer Academic Publishers, Netherlands, 1992.
[3] Nakib, R., Frangopol, D., Reliability-based structural optimization using interactive graphics, *Computers & Structures*, Vol. 37, pp. 27-34, 1990.
[4] Kishi, N., Chen, W.F., Goto, Y. & Matsuoka, K.G., Design aid of semi-rigid Connections for frame analysis, *Engineering Journal, AISC*, Vol. 30, No. 3, pp. 90-107, 1993.
[5] Hadianfard, M.A., Razani, R., Effects of semi-rigid behavior of connections in the reliability of steel frames, *Structural Safety*, 25, pp.123-138, 2003.
[6] Thoft-Christensen, P., Murotsu, Y., *Application of Structural System Reliability theory*, Springer-Verlag, Berlin, Heidelberg, 1986.
[7] Chen, W.F., Lui, E.M., *Stability Design of Steel Frames*, CRC Press, Inc., 1991.

Displacement maximisation of haunched-rafter Pitched-roof Steel Portal Frames

H. K. Issa & F. A. Mohammad
School of Architecture, Design and the Built Environment, Nottingham Trent University, UK

Abstract

Pitched-roof Steel Portal Frames (SPFs) are common structures used in single storey buildings. It is necessary for this popular steelwork to pass through an optimisation process to minimise the total cost of the frame. Optimisation in terms of weight is well documented with different optimisation techniques. However, this approach to displacement maximisation is somehow rare. In this paper, an attempt is made to perform optimisation in terms of lateral displacements. To achieve this, a modified distributed genetic algorithm (DGA) is used to maximise lateral displacements of the SPF while the constraints meet the requirements of BS 5950. Furthermore, software 'DO-DGA' (Design Optimisation using DGA), coded by Visual Basic 6.0, has been developed by the authors to loop the optimisation process. Although the appearance and form of SPFs are simple, according to BS 5950, there are more limitations to be checked than in complex structures. A stiffness matrix has been derived for the haunched part of the rafter using a column analogy and a virtual work method to involve this part in the analysis process. Through two benchmark examples, the comparison is made between the results of weight minimisation and displacement maximisation.
Keywords: distributed genetic algorithm, steel portal frame, optimisation, displacement maximisation.

1 Introduction

Single storey buildings are widely used in the UK; it is estimated that 50% of the single storey steel work buildings are constructed by steel portal frames [12]. Because of its economy and versatility for large spans in the construction of

WIT Transactions on The Built Environment, Vol 106, © 2009 WIT Press
www.witpress.com, ISSN 1743-3509 (on-line)
doi:10.2495/OP090131

pitched roofs such as shopping centres, warehouses, retail shops, pools, factories, etc, the steel portal frame has become the most often used structure within this sector.

Any structural designer attempts to conduct an economical design. This can be achieved by formulating a design problem and solving it by an optimisation technique while meeting the requirements of a code of practice to control the safety of the structure [13]. However, due to the large number of iterations in implementing the optimisation technique, it cannot be achieved by using the designer's experiences and intuition. As it is believed that the major cost of structural steelwork is its own weight, approaches to minimising the weight have become increasingly interesting for researchers. However, it necessitates choosing different approaches for those types of structure that are controlled by deflection and displacement. Minimising the weight is relatively well documented in literatures (see, for example, [1, 4, 6, 7, 10, 13]. To the knowledge of the authors there has not been any attempt to investigate the displacement maximisation of the steel structure thus far. In this paper, a modified distributed genetic algorithm (DGA) is implemented to conduct both weight minimisation and displacement maximisation. Through two benchmark examples, the comparison is made to portray the more effective approach for the design optimisation.

2 Distributed genetic algorithm

The basic mechanics of the Genetic Algorithm (GA) are based on randomised procedures of the selection and reproduction of the population of individuals and copying the fittest individuals into the next generation. A GA moves from one generation to another until either a certain individual dominated population or a predetermined maximum number of generations are reached. A basic GA consists of three main operators; reproduction, crossover and mutation. In the reproduction stage, a set of the population are selected for mating depending on their fitness values, which represent the objective function. If any constraint is violated, a penalty is applied to the objective function. The value of the penalty is related to the degree to which the constraints are violated.

Camp *et al.* [2] used three crossover schemes, fixed, flexible, and uniform, to minimise the weight of the structure. The crossover and mutation probabilities were 0.85 and 0.05 and the maximum crossover point was 3. Kameshki and Saka [6] applied a GA for the optimum design of unbraced multi-storey frames with semi-rigid beam-to-column connection. They adopted constant values of 0.001 for mutation probability. Toropov and Mahfouz [13] modified the GA to improve its rate of convergence. The modified GA was linked to a system of structural design rules, interacting with a finite element package in order to obtain minimum weight designs of plane structural steel frames. Saka [11] studied the optimum design of pitched roof steel portal frames using the GA. He used only gravity loads to minimise the weight of the steel portal frame. Degertekin *et al.* [3] implemented the GA to investigate the optimal load and resistance factor design. They applied a uniform crossover and a simple mutation probability with values of 0.001 and 0.002, as they found it suitable in the examples. The elapsed time for the process was 21.2min while 4512 frame analysis was required (6 design variables).

In the DGA, the performance of the conventional GA is improved by some minor modifications in its main algorithm that leads to quicker convergence and higher searching capability compared to the conventional GA [9]. Adopting the migration idea of the population, the DGA uses a number of population groups and implements genetic operations in parallel for all populations existing in different groups. Then, the best populations of each group migrate to another group, making it possible for them to contribute in quicker converge than the GA into an optimum solution.

In this research, a DGA has been modified for the purpose of improving the algorithm performance and saving the computation time in convergence into an optimum solution. Furthermore, an attempt is made to express the design problems in terms of weight minimisation and displacement maximisation to articulate and highlight the efficiency of both objective functions. For this purpose a software coded by Visual Basic 6.0, called Design Optimisation with DGA (DO-DGA), has been developed to conduct the optimisation process. The major aspects of modification that DO-DGA follows are stated as the following:

- In addition to other crossover schemes defined by literature review, a four-point crossover has been added to the crossover scheme.
- Although twin is not meaningful in genetic algorithm, DO-DGA had adopted this idea to produce more offspring. As the best parents in the population can give better offspring, a probability has been assigned to the parents that allows them to undergo crossover operation twice, resulting in producing more offspring. This will make it possible to increase the number of better individuals among the population of the group.
- The elite individuals are included in genetic operations while they are already secured to be dropped into the next generation. The elite individuals are participating in reproduction twice to increase the existence of fit individuals among the population.
- Similar individuals in the population do not undergo structural analysis and fitness calculation as many times as they are. Because of similarity in the population, the process is conducted once and the fitness value is assigned to all similar individuals at the same time.
- In contrast to the literatures which have addressed a constant value for mutation probability, DO-DGA uses a varied mutation probability. This assists the algorithms to make more diversity among population and consequently more feasible design spaces can be employed to reach the fittest individuals. Reaching to global optimum requires best diversity among population. The mutation probability can be formulated as follows:

$$P_m^{G_C} = P_m^{\max} - \frac{e^{-1/20} - e^{-G_C/20}}{e^{-1/20} - e^{-N_G/20}}(P_m^{\max} - P_m^{\min}) \tag{1}$$

where P_m^{Gc}= mutation probability of the current generation; P_m^{max} and P_m^{min}= maximum and minimum mutation probabilities; G_C = number of current generation; and N_G = number of predetermined generations.

- The algorithm uses a DO-DGA penalty function when the aim is to minimise the weight as below:

$$C = \begin{cases} 0 & \text{If } g_i \le 0 \\ g_i & \text{If } 0 < g_i \le 1 \\ 2g_i & \text{If } g_i > 1 \end{cases} \tag{2}$$

Whereby the fitness function is defined as follows:

$$F = W(1 + C) \tag{3}$$

Whereas a simple yet 'exact' penalty function [13] is adopted for maximising the lateral displacement.

$$F = \begin{cases} \sum \delta u & \text{If } g_i \le 0 \\ 0 & \text{If } g_i > 0 \end{cases} \tag{4}$$

where, g_i = value of constraint i; F = fitness value; W = total weight of frame; $\sum\delta u$ = sum of lateral displacement; and C = penalty value.

In design optimisation using genetic algorithm, the individuals that violate the constraints are not commonly discarded as it is believed that those individuals may produce better offspring after crossover and mutation operations. However, it is necessary to discard them when the aim is to maximise the lateral displacement. This is because the fitness function, in this case, depends not only on area and length (volume) but also on moment of inertia (stiffness). As there is no one-to-one relationship between the area and the moment of inertia of the section, formulating a comprehensive fitness equation is somehow impossible, because the bigger area for the section does not imply to have bigger moment of inertia and vice versa. As a result, lateral displacement of the bigger area may be larger than the lateral displacement of smaller area.

In addition to those modifications, DO-DGA uses a general stiffness matrix for both prismatic and non-prismatic members which will be explained later.

3 Analysis

As long as the direct stiffness method is used to analyse the structure, it is required to form stiffness matrices for structural elements and add them up to reach a global stiffness matrix for whole structure. The stiffness matrix for prismatic member is well documented through the text books. On the other hand the derived stiffness matrices for non-prismatic members are time-consuming as a subroutine must be formed to calculate the integrals' equations of the matrix elements. Saka [11] implemented the stiffness matrix of non-prismatic member developed by Matheson *et al.* [cited in 11] to carry out the optimization process on steel portal frames. Luo *et al.* [8] adopted the transfer matrix method to a deduced general expression for the components of the stiffness matrix of non-prismatic members. Both stiffness matrices have been formed by some integration components. In this paper a virtual work method and a column analogy [5] is used to derive a general stiffness matrix for the structural elements. The virtual work was implemented to derive the axial stiffness coefficient, whereas a column analogy was employed to derive the non-prismatic stiffness matrix for bending and shear effects. Accordingly, a member's stiffness matrix may have the form below:

$$[k]=\begin{bmatrix} \frac{E}{2L}(A_1+A_2) & 0 & 0 & -\frac{E}{2L}(A_1+A_2) & 0 & 0 \\ & \frac{1}{\int \frac{x^2}{EI_x}dx} & \frac{L_1}{\int \frac{x^2}{EI_x}dx} & 0 & -\frac{1}{\int \frac{x^2}{EI_x}dx} & \frac{L_2}{\int \frac{x^2}{EI_x}dx} \\ & & \left(\frac{1}{\int \frac{1}{EI_x}dx}+\frac{L_1^2}{\int \frac{x^2}{EI_x}dx}\right) & 0 & -\frac{L_1}{\int \frac{x^2}{EI_x}dx} & \left(-\frac{1}{\int \frac{1}{EI_x}dx}+\frac{L_1L_2}{\int \frac{x^2}{EI_x}dx}\right) \\ & & & \frac{E}{2L}(A_1+A_2) & 0 & 0 \\ & Symmetric & & & \frac{1}{\int \frac{x^2}{EI_x}dx} & -\frac{L_2}{\int \frac{x^2}{EI_x}dx} \\ & & & & & \left(\frac{1}{\int \frac{1}{EI_x}dx}+\frac{L_2^2}{\int \frac{x^2}{EI_x}dx}\right) \end{bmatrix} \tag{5}$$

where k = stiffness matrix of single member; E = modulus of elasticity; I_x = varied moment of inertia; $L_1$and L_2 = distances from member ends to the elastic centre; A_1 and A_2 = areas of member ends.

To eliminate the integrations in matrix elements, the member's stiffness matrix is passed through regression analysis whereby an additional subroutine is not required to code for the integration. Entire eighty standard steel sections available in BS 5950 are engaged into regression analysis. After immense regression analysis and data collection the stiffness matrix can be refined as below.

When A1 $\geq$ A2

$$[k]=\frac{E}{L^2}\begin{bmatrix} 0.50A'L & 0 & 0 & -0.50A'L & 0 & 0 \\ & 0.32(12I'/L) & 0.36(6I') & 0 & -0.32(12I'/L) & 0.29(6I') \\ & & 0.39(4I'L) & 0 & -0.36(6I') & 0.28(2I'L) \\ & & & 0.50(A'L) & 0 & 0 \\ & & & & 0.32(12I'/L) & -0.29(6I') \\ & & & & & 0.29(4I'L) \end{bmatrix} \tag{6}$$

where, I' = sum of moments of inertia of member ends; A' = sum of areas of member ends; L = length of the member.

For the prismatic member where the difference between the depths of member ends is zero, the stiffness matrix will have the form of conventional one given in any structure analysis textbooks.

4 Optimum design to BS 5950

BS 5950 states that when an elastic analysis is used for the design of steel framework such as the one shown in fig. 1, the capacity and buckling resistance should be calculated. It is required to use the effective length equal to that between two intermediate restraints.

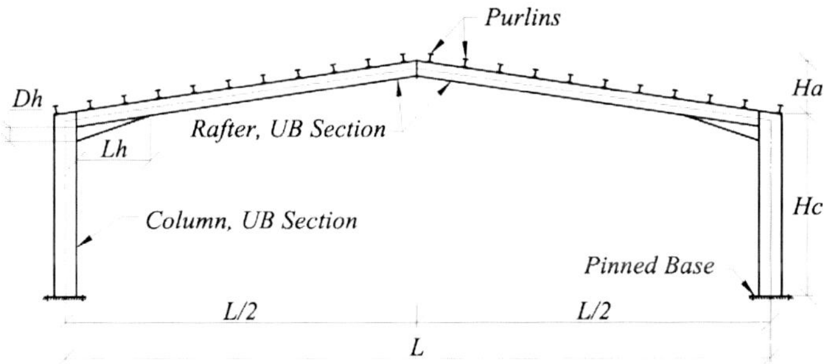

Figure 1: Typical pitched roof steel portal frame.

In the design of pitched roof steel portal frames, it is common to have the same universal beam section for both rafters and a different universal beam sections for the columns. For the reason of economy, the same section of rafter is used to produce the haunch. Therefore, the optimum design of the pitched roof steel portal frame necessitates using two design variables; one for rafter and its haunch and another for the columns. However, if it is necessary to use different section for the haunched section, the number of design variables increases to three.

The design of pitched roof steel portal frame with haunched eaves, when the objective is obtaining minimum weight and maximum displacement, and the constraints are implemented according to BS 5950 has the following form of formula:

$$\text{Minimise } W = \sum_{i=1}^{ng} \gamma_{im} \sum_{j=1}^{nm} V_j \tag{7}$$

$$\text{Maximise } \partial = \sum_{i=1}^{nj} \partial_{iu} \tag{8}$$

Subjected to:

$$\delta_i \le \delta_{iu} \qquad i=1, 2, 3 \ldots, nj \tag{9}$$

$$M_{xj} \leq M_{cxj} \qquad j=1,2, \ldots , nm \tag{10}$$

$$\frac{F_j}{A_{gj} p_y} + \frac{M_{xj}}{M_{cxj}} \leq 1 \qquad j=1, \ldots , nm \tag{11}$$

$$\frac{F_j}{A_{gj} p_{cj}} + \frac{m_j M_{xj}}{M_{bj}} \leq 1 j=1, \ldots , nm \tag{12}$$

$$\frac{F_j}{A_j} + \frac{M_j}{S_{xj}} \leq p_b \qquad j=1, 2, \ldots , nh \tag{13}$$

where W = total weight if frame; ng = number of member groups; γ_m = unit weigh of the member group; nm = number of members in a group; V_j = volume of member j; ∂ = total lateral displacements of joints; nj = total number of joints; ∂_{iu} = lateral displacement of joint i; δ_i = horizontal and vertical displacements of joint i, δ_{iu} = upper limit of displacements; F_j = axial member force of member j; A_{gj} = gross cross-sectional area of member j; p_y = design strength; M_{xj} = bending moment about major axis; M_{cxj} = bending moment capacity of member j; p_{cj} = compressive strength of member j; m_j = equivalent factor; M_{bj} = lateral torsional buckling resistance moment; S_{xj} = plastic section modulus; p_b = bending strength.

Eqn. (9) checks the displacement of the joints. BS 5950 has limited the horizontal displacement of the joints to column/300 and the upper limit of the beam deflection is span/360. Eqn. (10) checks the moment capacity of the member section-section. Eqn. (11) defines the load capacity check for beam-column with semi-compact or slender cross section. Eqn. (12) is the simplified approach of the overall buckling check for beam-column. Eqn. (13) should be checked at any point of the haunched rafter.

As the nature of the structural optimisation variables are discrete, the solution of the optimum design problem given in eqns. (7) and (8) necessitates selecting universal beam section from the table of standard section for rafters, columns and haunched section.

5 Benchmark examples

Two examples are used to illustrate the efficiency of both defined objective functions. The steel portal frames given as examples, are assumed to experience different loads combinations. In addition to the imposed load, one of the frames is subjected to wind load which act only on one of the rafter. Unlike what was done by Saka [11], the structural members are not subdivided into smaller elements whereas the structure is summarised into six members. Furthermore, it is decided to apply seven genes for each design variables to include all eighty standard steel sections into calculation. The variables that exceed from the upper limit of steel section number (80) are altered to a random number between 0 and 79. The steel grade of the sections is S275 and 3 lateral restraints have been used for columns. The length and depth of the haunched part of the rafter are varied to find the optimum size of the haunched member. The length of the haunched part varies between 0.05m and 6.40 increasing by 0.05m whereas the depth of the haunched part varies between 0.01m and 1.28m increasing by 0.01m. As a result, the number of point loads is changed on the haunched part and the rest of the rafter. DO-DGA has handled this changing by adding a subroutine for load transfer. The number of population for each group is 26. This was decided after a number of trials that was done to choose the minimum number which gives the best solution. The maximum number of generations is 100 which the algorithm terminates in a certain generation due to domination of a specific individual. A 0.85 value is adopted for the crossover probability. The maximum mutation probability is 0.20 and the minimum value is 0.0005. 30% of the population are secured as elite for the next generation. Simultaneously, 30% of the fit population are chosen to migrate to the other group with the migration interval of 3 generations. A 0.40 value is assigned for the probability of twice crossover to produce double offspring (twins). As there is no guarantee that the GA can give the optimum solution, the program is performed 10 times, represented as the number of the 'seed'. At the beginning the decided number of run was 20. However, as the optimum result gained within 20 runs was the same as 10 runs, therefore, the number of runs was fixed to 10 for reducing the computation time.

5.1 Frame with self-weight and imposed load

5.1.1 Weight minimisation

An imposed load of 5kN, in addition to the frame's self-weight, is applied to the pitched roof steel portal frame with the span of 20.0m, the column height of 5.0m and the overall height of 6.5m (as shown in fig 2). To design the frame, two groups of population are employed. The best individuals of the first group migrate to the first group to enhance the quality of the first group population. Both length and depth of the haunched part are varied according to the decided range.

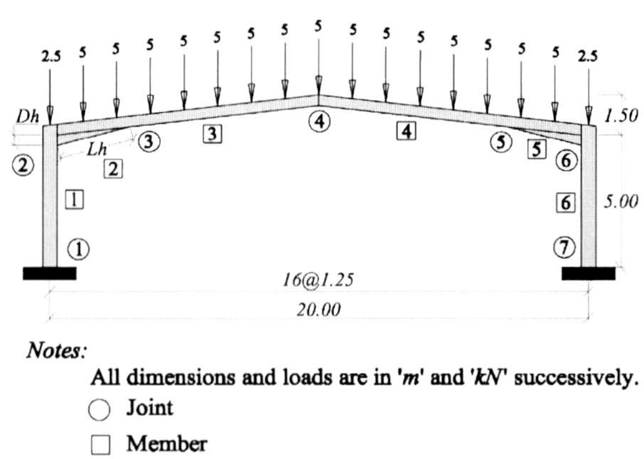

Figure 2: The pitched-roof SPF subject to imposed load and self-weight.

The optimum solutions gained after ten runs are given in table 1.

Browsing the results, the consistency of selecting the steel sections for rafters and columns is remarkable. As the range of the depth and length are quite big with smaller step, DO-DGA has found different haunch length and depth at each run. The weights of the optimum solution for ten runs are close to each other showing the efficiency of the program. The optimum gained weight is 1133.9 kg as shown in runs 3, 7, and 9. The optimum haunch length and depth are 1.60m and 0.64m respectively. The optimum design was achieved within 43 generations which last 194 seconds. The constraint associated to interaction of the axial force and the moment reached to 0.94 whereas the vertical displacement of the apex reached to its maximum value (55.5mm) due to large value of live load.

Table 1: Result of weight minimisation.

Run	Column section	Rafter section	Haunch length	Haunch depth	Weight , kg	Gene-ration
1	406x140x39	356x127x33	1.60	0.69	1137.7	46
2	406x140x39	356x127x33	1.60	0.67	1136.2	46
3	406x140x39	356x127x33	1.60	0.64	1133.9	49
4	406x140x39	356x127x33	1.75	0.58	1135.9	47
5	406x140x39	356x127x33	1.65	0.65	1137.0	55
6	406x140x39	356x127x33	1.75	0.58	1135.9	47
7	406x140x39	356x127x33	1.60	0.64	1133.9	43
8	406x140x39	356x127x33	1.80	0.56	1136.4	49
9	406x140x39	356x127x33	1.60	0.64	1133.9	46
10	406x140x39	356x127x33	1.45	0.83	1139.9	43

However, varied depth necessitates building up an extra flange and web, and then welding them to the rafter which results in increased time and cost. It is recommended for haunch in eaves, besides using the same cross-section as rafter, to use the same depth as rafter. This is done in practice by cutting the section equal to the obtained optimum haunch length and halving it diagonally and longitudinally to avoid wasting the steel material. Having this idea, the optimum obtained weight was slightly heavier than using the varied depth. It increased the weight by 0.3% whereas the optimum haunch length was 2.40m. The convergence took place after 31 generation within 123 seconds time.

5.1.2 Displacement maximisation

In contrast to the weight minimisation, displacement maximisation does not assign consistent member sections to the rafters and columns. Furthermore, the optimum solution only happened once within ten runs. The reason refers to penalty function; as the violent individuals do not contribute in the crossover and mutation operations. The results show a bigger depth and smaller length of the haunch (Table 2) comparing to the weight minimisation. The optimum solution took place after 51 generation which has a weight of 1255.94 kg. It is observed that the interaction limit reached to 1.00 whereas the lateral displacement did not reach the upper limit.

In the case when the depth is fixed, the optimum solution is lighter than when a varied depth is the design variable. The weight reduced by 5% and the optimum solution reached after 49 generations with a haunch length of 0.70m while the interaction limit reached to 0.99 and lateral displacement did not reach its upper limit.

Table 2: Result of lateral displacement maximisation.

Run	Column section	Rafter section	Haunch length	Haunch depth	Weight Kg	Gene-ration
1	406x178x54	305x127x48	0.05	0.89	1518.2	39
2	305x165x54	305x165x54	0.05	0.55	1635.5	52
3	406x178x54	305x127x48	0.10	0.54	1520.2	54
4	406x140x46	356x127x39	0.15	0.27	1255.9	51
5	406x178x67	305x127x42	0.55	0.27	1540.3	48
6	406x140x46	305x127x48	0.30	0.71	1455.8	42
7	356x171x57	305x165x46	0.15	0.66	1512.0	53
8	406x178x74	305x127x42	0.35	0.21	1602.0	56
9	406x178x60	305x127x42	0.40	1.07	1484.4	53
10	356x171x57	356x127x39	0.05	0.98	1364.3	54

5.2 Frame with dead and wind loads

The second example is shown in fig 3. A combination of dead load with a value of 9.8 kN generated from purlins and wind load with a value of 2.8 kN are applied to the frame. Only one side of the frame is experienced by wind load whereas whole rafter of the frame is subject to the dead load. A notional horizontal force with a value of 0.049kN is applied to node 2. The span of the frame is 24m and the column and overall heights are 5.5m and 7.0m respectively. The same genetic parameters are used for this example.

5.2.1 Weight minimisation

The optimum length of the haunch has reached its upper range which is 6.40m whereas the depth of haunch is quite small. The frame had a weight of 2557.6 kg at its optimum solution. 70% of the runs are the same section and haunch dimensions. The optimum solution was obtained after 46 generations. Unlike the displacements, the interaction limitation reached to its upper limit. The convergence lasted 156 seconds.

Taking the depth of haunch fixed, ended up with an 18% heavier frame while the interaction constraints control the design. The optimum obtained length was 1.60m.

5.2.2 Displacement maximisation

The obtained optimum design had an overall weight of 3315.1 kg which is 29% heavier than when the weight is minimised. In addition, the optimum solution does not take place consistently within 10 runs of the program. The convergence to optimum solution took place after 51 generations with last 231 seconds time.

Once again, the interaction constraints control over the optimum design while the lateral displacements do not reach upper limit. However, the values of lateral displacements increased by 4% comparing to the weight minimisation.

Having the results considering fixed depth of the haunch, the optimum design comes up with slightly heavier weight than considering the varied depth by 0.4%.

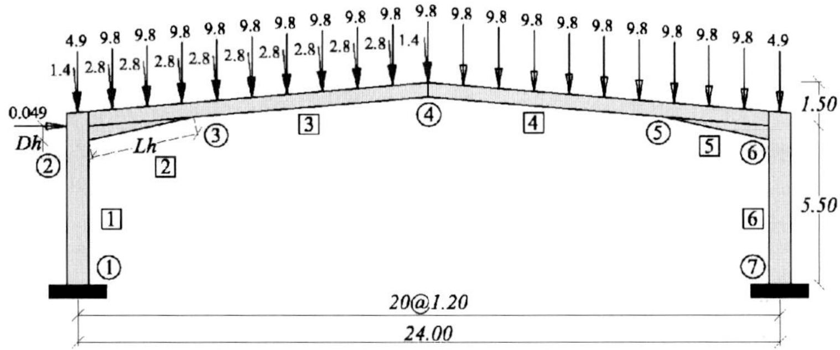

Notes:
All dimensions and loads are in '*m*' and '*kN*' successively.

Figure 3: Pitched-roof SPF experiences dead and wind loads.

6 Conclusion

A modified distributed genetic algorithm was used to enhance the speed of convergence into design solution. The modification includes using a varied mutation probability and bringing the elite individuals into the genetic operations twice while they are already secured to be dropped into the next generation. The mutation scheme assisted the algorithm to diversify the population in the earlier generation, due to high probability value, and making the design possible to access the feasible design space as much as possible. At later generations the probability value slipped down and allowed the problem to converge into the optimum solution, which happened within smaller number of generation, leading to a substantial decrease in computation time. The general stiffness matrix was another factor to reduce the computation time as it eliminated more loops for the integration for non-prismatic members. Using different mutation, reproduction schemes, fitness value calculation, DO-DGA can obtain the optimum solution within a few minutes so that it can be brought into daily office-use by structural engineers. Two different objective functions were used to appraise their efficiency. Two benchmark examples revealed that the minimisation of weight gives better results than maximisation lateral displacement does.

References

[1] Balling, R.J., Briggs, R.R., & Gillman, K., Multiple optimum size/shape/topology designs for skeletal structures using a genetic

algorithm. *ASCE Journal of Structural Engineering,* **132 (7)**, pp. 1158 – 1165, 2006.

[2] Camp, C., Pezeshk, S., & Cao G., Optimum design of two-dimensional structures using a genetic algorithm. *ASCE Journal of Structural Engineering*, **124 (5)**, pp. 551 – 559, 1998.

[3] Degertekin, S.O., Saka, M.P., & Hayalioglu, M.S., Optimal load and resistance factor design of geometrically nonlinear steel space frames via tabu search and genetic algorithm. *Engineering Structures*, **30**, pp. 197-205, 2008.

[4] Foley, C., & Schinler, D., Automated design of steel frames using advanced analysis and object-oriented evolutionary computation. *ASCE Journal of Structural Engineering,* **129 (5)**, pp. 648 – 660, 2003.

[5] Ghali, A., Neville, A.M., & Brown, T.G., *Structural Analysis; A Unified Classical and Matrix Approach,* Spon Press: New York, 2003.

[6] Kameshki, E.S., & Saka, M.P., Optimum design of non-linear steel frames with semi-rigid connection using a genetic algorithm. *Computer and Structures*, **79**, pp. 1593-1604, 2001.

[7] Liu, X., Yi, W.J., Li, Q.S., Shen, P.S., Genetic evolutionary structural optimization. *Journal of Constructional Steel Research,* **5**, pp. 346-358, 2007.

[8] Luo, Y.Z., Xu, X., & Wu, F. Accurate stiffness matrix for non-prismatic members. ASCE, *Journal of Structural Engineering*, **133(8)**, pp. 1168-1175, 2007.

[9] Mühlenbein, H., Schomisch, M., & Born, J., The parallel genetic algorithms as a function optimizer. *Parallel Computing*, **17**, pp. 619–632, 1991.

[10] Pezeshk, S., Camp, C.V., & Chen, D., Design of non-linear framed structures using genetic optimization. *ASCE Journal of Structural Engineering*, **126 (3)**, pp. 382 – 388, 2000.

[11] Saka, M.P., Optimum design of pitched roof steel frames with haunched rafter by genetic algorithms. *Computers and Structures*, **81**, pp. 1967 – 1978, 2003.

[12] Salter, P.R., Malik, A.S., & King, C.M., ed., Design of single-span steel portal frames to BS 5950-1:2000. The steel Construction Institute: Berkshire, 2004.

[13] Toropov, V.V., & Mahfouz, S.Y., Design optimization of structural steelwork using a genetic algorithm, FEM and a system of design rules. *Engineering Computations,* **18 (3/4)**, pp. 437-459, 2001.

Optimal bracing type and position to minimize lateral drift in high-rise buildings

O. Alshamrani[1], G. G. Schierle[2], K. Galal[1] & D. Vergun[2]
[1]*Department of Building, Civil, and Environmental Engineering, Concordia University, Montreal, Canada*
[2]*USC, School of Architecture, Los Angeles, USA*

Abstract

Selection of high-rise bracing type and position is based on many factors: their degree of resistance to lateral forces, cost, complexity of connection and assembly, and the integration with Architectural objectives. This paper includes selection of the optimal bracing type and position for a high-raise office building in Dammam Saudi Arabia. This 900 square meters per floor prototype is used as case study to compare four bracing types: single diagonal bracing, chevron shape bracing, cross (X) bracing and eccentric bracing; as well as two positions: bracing at the core and external parameter bracing (building facade). The case study building height is 40 stories in an area of high wind velocity (145 km/h) where no critical seismic force is recorded. The STAAD Pro 2005 software is used to analyze these systems according to allowable stress requirements for an objective function to minimize drift for wind speed of 145 kilometers per hour. The result of this study shows that adding bracing members to the moment frame structure increases stability and reduces drift. Chevron bracing and cross bracing have the highest resistance to the lateral drift compared to the others; but cross bracing is more costly due to more joints. Therefore chevron is the optimal bracing type. Also this study shows that adding the braces to the core of building reduces the drift much more than adding them to the building facades.

Keywords: optimal, lateral drift, cross bracing, diagonal bracing, eccentric bracing, chevron bracing stability, stiffness.

1 Introduction

Selection of optimal bracing type and position in high rise buildings is a very significant factor to control the lateral drift caused by lateral load. Controlling

WIT Transactions on The Built Environment, Vol 106, © 2009 WIT Press
www.witpress.com, ISSN 1743-3509 (on-line)
doi:10.2495/OP090141

the lateral drift is very essential to provide comfort to human beings and avoid sickness motion. In addition, large drift may cause danger to life and cause façade failure, or even might cause building collapse. Since wind load is more critical than seismic load in Saudi Arabia; this paper investigates wind effect on high-rise Steel buildings. Since wind load increases with height and increases lateral drift, decreasing lateral drift is a vital criterion to select bracing type and position. Basically, this paper investigates the reduction of lateral drift in high-rise buildings, nevertheless considering the integration with the architectural objectives, which governs the optimal selection of the bracing type and position in high-rise buildings.

2 Background

Braced frames resist gravity load in beam bending and column compression; and lateral load in axial compression and tension [1]. Braced frames are stiffer than moment frames, which is good in areas of wind load but it will increase seismic forces. However, braced frames are not as rigid as shear wall systems. There are several types of braced frame like single diagonal, chevron shape brace, V shape brace, cross (X) bracing, and K bracing [1].

X bracing is considered as stiffer than others, which increase the rigidity of the building. On the other hand, it has more connection joint that will increase the construction cost. Selecting the type of braces depends on their position in the building. Also, it depends on the function of the place. For example, if the space has openings like doors or windows one may select chevron shape bracing to allow openings between braces. However, some architects like to put the braced frame in the external frame to give expression to the building while others like to put it in the internal core of the building [4].

Axial action of braces resists the lateral shear caused by lateral load. Hence, it is an efficient system for steel structures to resist the lateral force and provide strength and stiffness as well[2]. Because the lateral force is reversible, the axial force acting in columns, girders, bracings also reverse. Hence, the braces must act in both compression and tension, but it must be so efficient to resist the compression force [2]. To resist the lateral shear the girders are in compression while the braces are in tension or compression. Compression in leeward columns and tension in windward columns resist the external moment. The flexural deformation in structure is caused by the shrinking of the leeward columns and the expansion of the windward columns [2]. The high-rise buildings deflect predominantly in flexural mode while the low-rise buildings deflect predominantly shear mode [2].

3 Methodology

Selection of optimal bracing type and position process is shown in figure 1. Four bracing types: (single diagonal bracing, (X) cross bracing, chevron shape bracing, and eccentric bracing) are tested in this study; as well as two positions: bracing at the core and external parameter bracing (building facade). This process implies eight combinations of bracing types and position for 40 story building height. Each bracing's type and position is passed through criteria process to minimize lateral drift (main criterion) as well as integration with architectural objectives (secondary criterion). All bracing types and positions combinations are entered to the design evaluation of STAAD Pro 2005. By comparing all result to the steel moment frame system, the optimal selection of the bracing type and position is verified. Accordingly, minimal lateral drift is achieved.

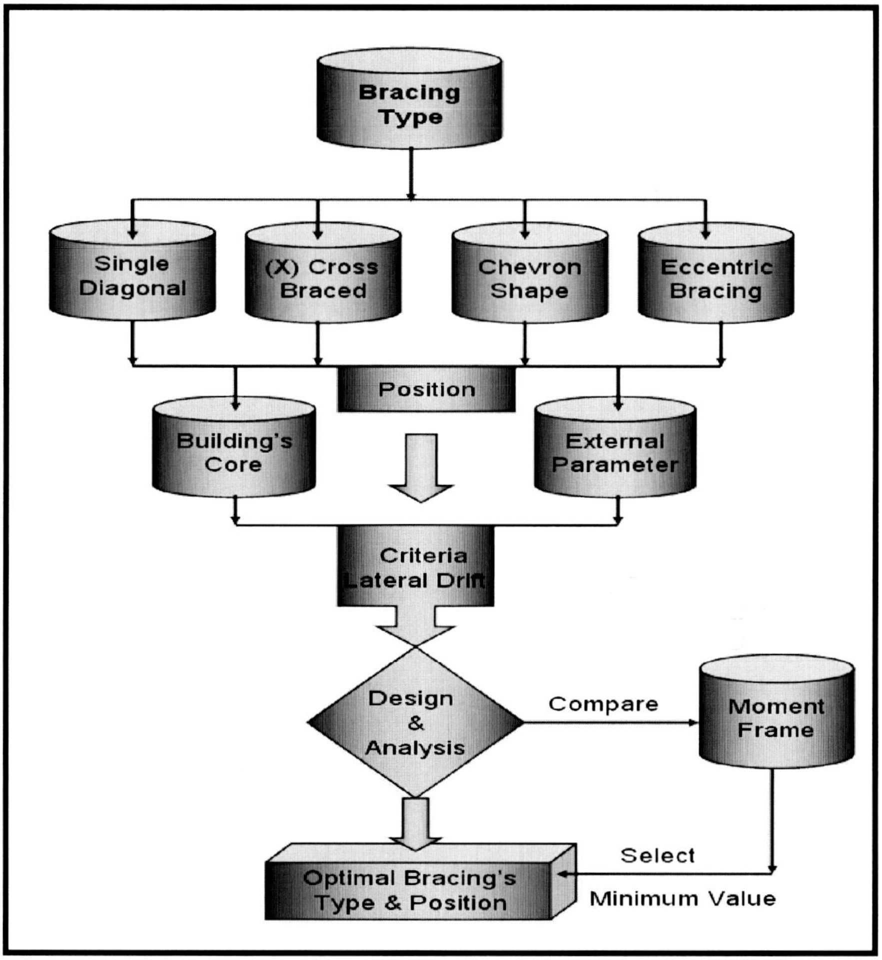

Figure 1: Selection of the optimal bracing's type and position process.

4 Assumptions

Using the structural design and analysis software (STAAD Pro), the case study assumed: International Building Code IBC 03; 145 kilometre/hour maximum wind speed; allowable stress design method for schematic structural design; material strength of steel: 50 ksi. Structural members are designed to meet the allowable stresses (30 ksi steel beam, 25 ksi steel columns and bracings due to column buckling) and maximum allowable lateral drift (h/200). Gravity load was assumed applied simultaneously with wind load, assuming combined dead and live loads uniform distributed loads of 1.7 kip/ft. Lateral load per level was assumed as shown in Fig. 3 acting on wind and lee sides in Kips.

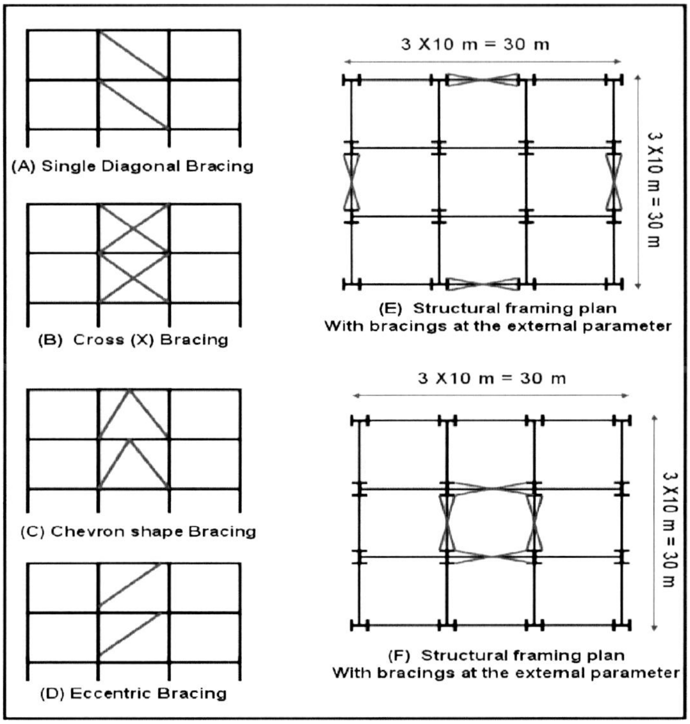

Figure 2: Prototype plans and sections for different bracing types and positions.

Variables investigated in the case study:

- Bracing types (single diagonal, chevron shape, (X) bracing and eccentric bracing) as shown in Fig. 2.
- Bracing positions (at the core and external parameter (building façade)).

The lateral drift is measured at each level after defining member size for strength to assure the actual lateral drift is less than the maximum allowable. The tested case study consists of 3 bay frames with 33x33 ft (10X10m) column spacing and 13 ft (4m) story height as shown in Fig. 2.

5 Wind load distribution

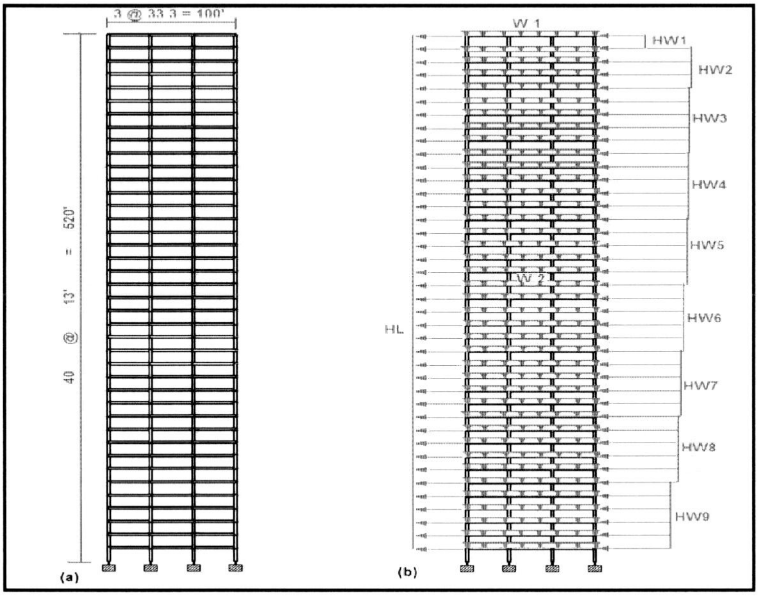

Figure 3: 40 story (520 ft) height prototype frame; Section (a), wind and gravity load (b).

The loads of the 40 story frame are shown at figure 3.

Gravity load:

Roof W 1 = 0.9 kip / ft
Floor W 2 = 1.7 kip / ft

Lateral Load:

Table 1: Wind loads for 40 story steel building.

Windward Side. (Pressure)		Leeward & Internal Pressure	
WH 1	5.85 kip	**Leeward Side. (Suction) HL**	**7.94 kip**
WH 2	11.69 kip		
WH 3	11.30 kip		
WH 4	10.92 kip		
WH 5	10.52 kip		
WH 6	10.00 kip	**Internal (Pressure) HL**	**3.97 kip**
WH 7	9.40 kip		
WH 8	8.60 kip		
WH 9	7.50 kip		

6 Steel members schedule

Table 2: Beams schedule for a 40 story steel braced frame.

Beam Size	Floor No.	Beam Size	Floor No.
W18 x 71	39,40	W18x 158	23-27
W18 x76	37,38	W18x 175	18 TO 22
W18 x 86	35,36	W18x 192	15-17
W18 x 97	33,34	W18x 158	11-14
W18 x 119	30,31,32	W18x 143	7 TO 10
W18x 130	28,29	W18x 130	1 TO 6

Table 3: Braces schedule for a 40 story steel braced frame.

Braces Size	Floor No.	Braces Size	Floor No.
W12 x 14	37 to 40	W12 x 35	17 to 21
W12 x 16	32 to 36	W12 x 50	13 to 16
W12 x 19	27 to 31	W12 x 65	8 to 12
W12 x 26	22 to 26	W12 x 79	1 to 7

Table 4: Columns schedule for a 40 story steel braced frame.

Column Size	Floor No.	Column Size	Floor No.
W14 x 193	37 TO 40	W14 x 665	15,16
W14 x 211	35,36	W14 x730	13,14
W14 x 283	33,34	W14 x825	11,12
W14 x 342	31,32	W14 x 1225	9,10
W14 x 370	29,30	W14 x 1305	7,8
W14 x 426	27,28	W14 x 1385	5,6
W14 x 500	23 TO 26	W14 x 1465	3,4
W14 x 550	21,22	W14 x 1545	1,2
W14 x 605	17 TO 20	-------------	---------

7 Bracing frame at the external parameter

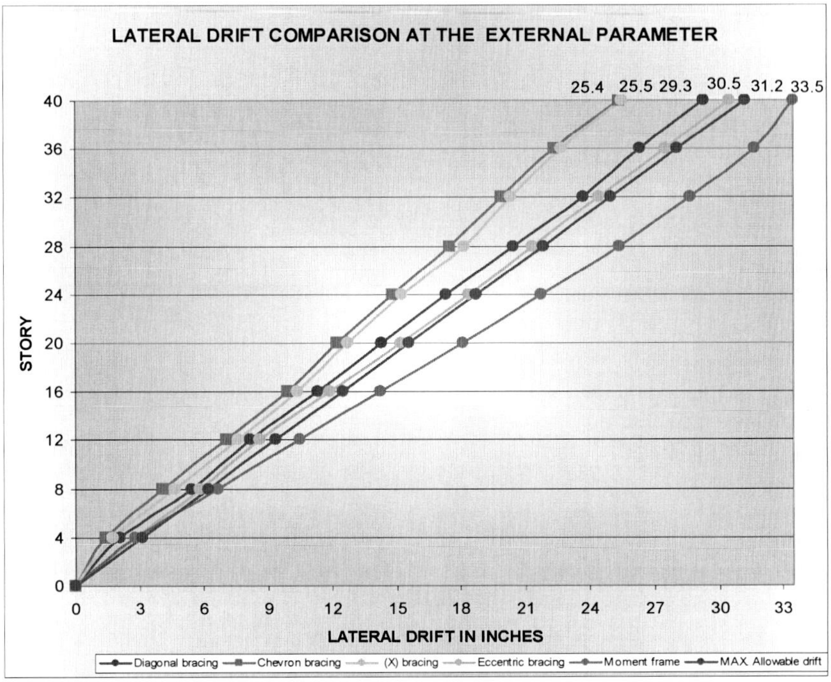

Figure 4: Lateral drift comparison for different bracing types at the external parameter (building façade).

Table 5: Lateral drift for different bracing types at the external parameter.

Floor Level	Diagonal bracing	Chevron bracing	(X) Cross bracing	Eccentric bracing	Moment frame	Allowable drift
4	2.12	1.41	1.75	2.81	2.79	3.12
8	5.41	4.11	4.65	5.80	6.66	6.24
12	8.22	7.07	7.61	8.62	10.47	9.36
16	11.31	9.89	10.34	11.91	14.22	12.48
20	14.27	12.21	12.72	15.20	18.06	15.60
24	17.32	14.78	15.21	18.40	21.76	18.72
28	20.45	17.51	18.15	21.33	25.35	21.84
32	23.67	19.87	20.31	24.41	28.66	24.96
36	26.34	22.35	22.70	27.48	31.66	28.08
40	29.30	25.40	25.50	30.50	33.50	31.20

8 Bracing frame at the core

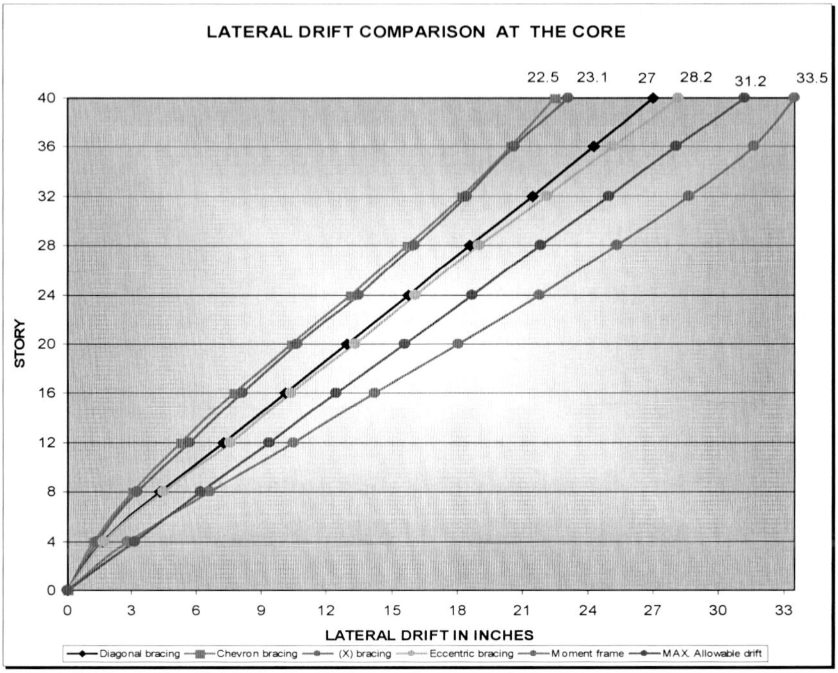

Figure 5: Lateral drift comparison for different bracing types at the core.

Table 6: Lateral drift for different bracing types at the core.

Floor Level	Diagonal bracing	Chevron bracing	(X)Cross bracing	Eccentric bracing	Moment frame	Allowable drift
4	1.65	1.23	1.4	1.70	2.79	3.12
8	4.32	3.05	3.25	4.45	6.66	6.24
12	7.32	5.28	5.7	7.60	10.47	9.36
16	10.1	7.78	8.5	10.4	14.22	12.48
20	12.93	10.41	10.65	13.3	18.06	15.60
24	15.77	13.1	13.5	16.1	21.76	18.72
28	18.60	15.72	16.3	19.03	25.35	21.84
32	21.44	18.22	18.42	22.10	28.66	24.96
36	24.27	20.46	20.6	25.18	31.66	28.08
40	27.11	22.5	23.1	28.20	33.55	31.2

9 Results

The result of this study shows that adding bracing members to the moment frame structure increases the stability and reduces drift. The maximum allowable drifts in x, z directions (side – sway) of 40 story building with (520 ft) height are 31.2 in. The allowable drift is estimated for the seismic load according to the International Building Code (IBC03) is 0.005 times building height. d = 520 ft x 0.005 x 12 = 31.2 in.

As shown in figure 4. The optimum bracing type at the external parameter (building façade) for a 40 story building is chevron shape bracing with lateral drift (25.4"). This bracing type minimizes lateral drift by 25% compared to the moment frame system

The second best bracing type at the external frame is (X) cross bracing with lateral drift (25.5"). It minimizes lateral drift by 24% compared to the moment frame system. The third best bracing type is a single diagonal bracing with lateral drift (29.3"). This type minimizes lateral drift by 13% compared to the moment frame system. The lowest recorded resistance to drift is provided by the eccentric bracing with lateral drift (30.5"). It minimizes lateral drift by 10% compared to the moment frame system as shown in fig. 4.

As shown in figures 5. The optimum bracing type at the core of building for a 40 story building is chevron shape bracing with lateral drift (22.5"). This bracing type minimizes lateral drift by 33% compared to the moment frame system.

The second best bracing type at the core of building is cross (X) bracing with lateral drift (23.07"). It minimizes lateral drift by 31% compared to the moment frame. The third best bracing type is a single diagonal bracing with lateral drift (27.1"). This type minimizes lateral drift by 20% compared to the moment frame system. The lowest effective bracing type is the eccentric bracing with lateral drift (28.2"). It minimizes lateral drift by 16% compared to the moment frame system as shown in fig. 5.

In addition, this study shows that by adding the braces to the building core, leads to the reduction of lateral drift much more than adding them to the building facades. For example, chevron shape bracing at the external frame recorded a lateral drift of (25.4"), which is 25% less than the moment frame system. On the other hand, chevron shape bracing at the core recorded a lateral drift of (22.5"), which is 33% less than the moment frame system as shown in figures 6, 7.

Furthermore, this study shows that chevron bracing and (X) cross bracing have the highest resistance to the lateral force compared to the others; but cross bracing is more costly due to more joints. Hence, chevron shape proves to be the optimal bracing. This optimization also is due to the flexibility of chevron bracing to the integration with architectural objectives criterion. For example, chevron shape bracing allows opening like doors, windows or other services passage while the (X) cross bracing could cause obstacles to some of theses openings and service passages.

This study also shows that 40 story building the total steel weight is 11.5 million lb for steel moment frame, and 9.8 million for steel braced frame.

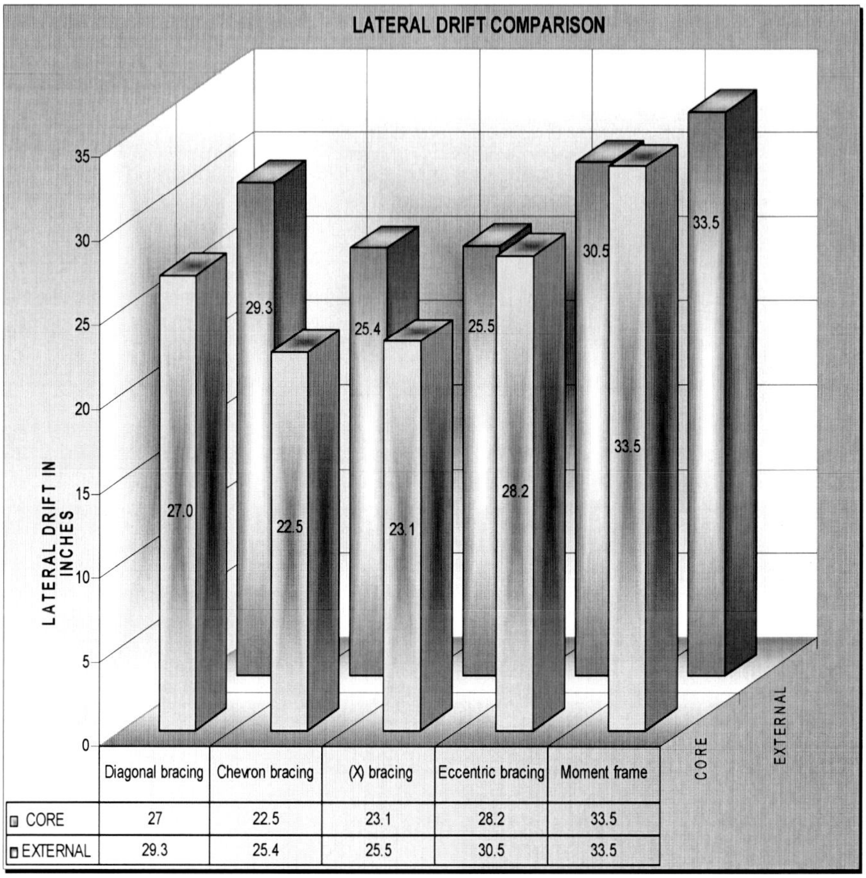

Figure 6: Lateral drift comparisons for different bracing types and positions.

Therefore, adding bracing to steel moment will reduce building mass 1.7 million lb which consequently will reduce the total cost.

10 Conclusion

The results of this study show that adding braced frame to steel moment frame building is important to reduce lateral drift as well as building weight and increase stability. The optimum bracing type whether at the facades or the building core for a 40 story building is chevron shape bracing. Also this study shows that adding the braces to the core of building reduce the drift much more than adding them to the facades. The eccentric braced frame is the most flexible braced system as proven in this study. Hence, it is a highly recommended type in seismic zone while it is not appropriate in high wind speed area. Chevron bracing and cross bracing have the highest resistance to the lateral drift compared to the others; but cross bracing is more costly due to more joints. Furthermore, chevron bracing is proved to be more flexible to the openings and services passages. Therefore the chevron shape is the optimal bracing type

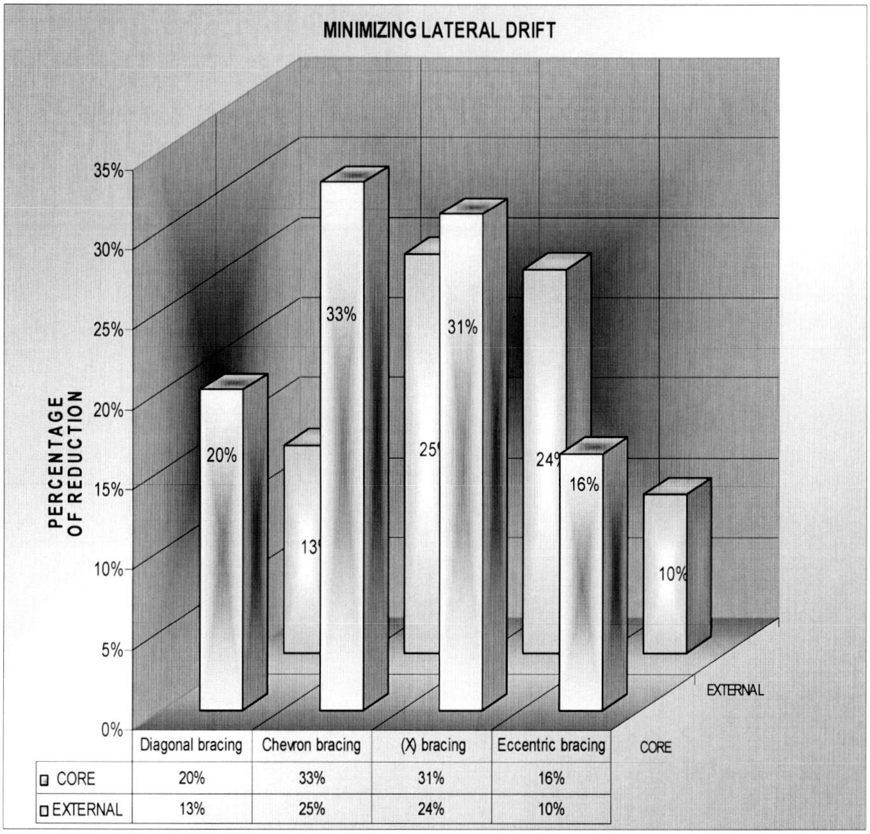

Figure 7: Minimizing lateral drift comparisons for different bracing types and positions.

References

[1] G. G Schierle, Structure and Design, University Readers, San Diego, 2008 http://www.universityreaders.com/titles/schierle/

[2] Ambrose, James and Vergun, Dimitry, 1995, Simplified Building Design for Wind and Earthquake Forces, University of Southern California, Third Edition, John Wiley & Sons, Inc: New York. ISBN 0-471-309968-3.

[3] Ambrose, James and Vergun, Dimitry, 1987, Design for Lateral Forces, University of Southern California, John Wiley & Sons, Inc: New York, ISBN 0-471-84889-1.

[4] Schierle, G. Goetz, 2006, Structures in Architecture, class notes, ARCH 613 University of Southern California.

[5] Blandra, T., 1993, Vibration of Buildings to Wind and Earthquake Loads, Springer-Verlag: London 1993, ISBN 0-387-19833

[6] IBC 03 International Building Code, Internal Code Council, Fall Church, Virginia, 2003.

[7] AISC Manual of Steel Construction Allowable Stress Design, American Institute of Steel Construction, Chicago, ninth edition, 1991.

[8] CRSI Concrete Reinforcing Steel Institute, Guide to Structural System Selection, 1997.

[9] Cheng-Yu Ho, Lateral Drift Cause by Wind Forces in High-Rise Steel Frame Master of Building Science Thesis, University of Southern California, August 1992.

[10] Cheng-Yu Ho and G. G Schierle, High-Rise Space Frame: Effect of Configuration on Lateral Drift, Proceedings, 4th International Conference on Space Structures, Surrey, Telford Publications.

Structural concept, static and dynamic properties of RPC-BSS with high durability

L. C. Jiang[1], R. Gao[1] & L. Li[2]
[1]*School of Civil Engineering, Beijing Jiaotong University, P.R. China*
[2]*Central Research Institute of Building and Construction, NCC Group, P.R. China*

Abstract

A new kind of beam string structure called RPC-BSS, which consists of reactive powder concrete upper chord and cables, is presented in this paper. The material of RPC provides superior mechanical properties and high durability, and the cable provides huge tensile force. The concept is based on the combination of the advantages of RPC and beam string technology. Furthermore, static and dynamic properties are discussed, and the influence of different factors is taken into consideration in the analysis. The analytical results show RPC-BSS is of longer span, less weight, better seismic characteristics and higher durability than conventional BSS.
Keywords: RPC-BSS, reactive powder concrete, high durability.

1 Introduction

Beam string structure (BSS) [1] is a kind of composite structure which is composed of a rigid member as the upper chord, a flexible member as the lower chord and some bracing bars as link. There are three familiar types of BSS. The first is tension beam, where the rigid member is a beam. The second called a tension arch, which is made up of a rigid arch as the upper chord. The third substitutes a two-way sloped beam for the arch.

The new material of RPC provides superior mechanical properties and high durability. The mechanical properties of RPC, including ultimate strength, Young's modulus, elastic response, ductility and capacity to absorb energy, etc, are better than those of HPC, SFRC and NSC. It is considered to be a

WIT Transactions on The Built Environment, Vol 106, © 2009 WIT Press
www.witpress.com, ISSN 1743-3509 (on-line)
doi:10.2495/OP090151

revolutionary material that provides a unique combination of ductility, strength, durability, and aesthetic flexibility.

2 Concept of RPC-BSS

To take full advantages of RPC and the mechanics of conventional BSS, a new kind of BSS with a box-section rigid member made of RPC is formed and named RPC-BSS. In RPC-BSS, the cable is prestressed to provide part of rigidity of the structure and bear the bracing bars, which act as elastic supports to offer resistance to deformation of the RPC member. The RPC member is mainly subjected to axial loads, at the same time subjected to some bending moment. The deformation of the structure could be controlled by adjusting the degree of prestress, and the value of tensile stress within the RPC members could be controlled too.

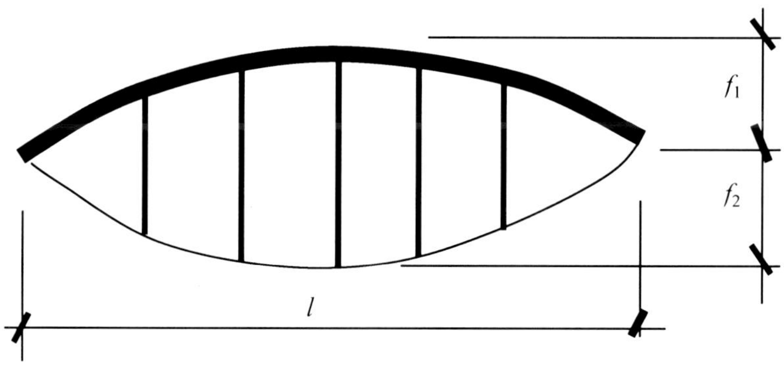

Figure 1: Model of RPC-BSS.

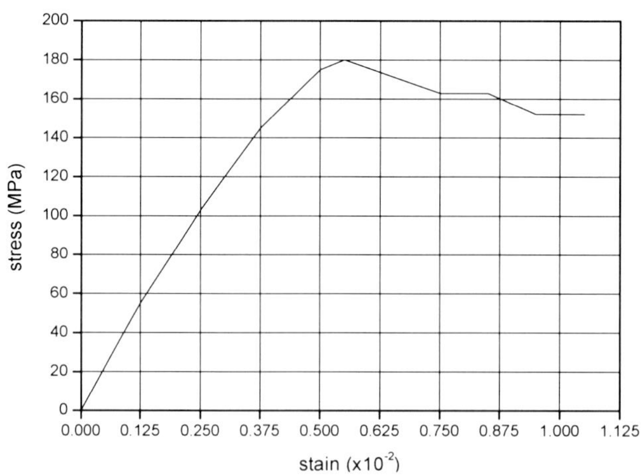

Figure 2: Strain-stress curve of RPC.

3 Static properties of RPC-BSS

During the whole process analysis, the finite element model of RPC-BSS is set up, in which beam elements and link elements are used to model the rigid upper chord, bracing bars and cables, as shown in Figure1. With the software ANSYS, displacement, stress, etc, are calculated and discussed. The curve of strain-stress of RPC [2] used in this calculation is shown in Figure 2.

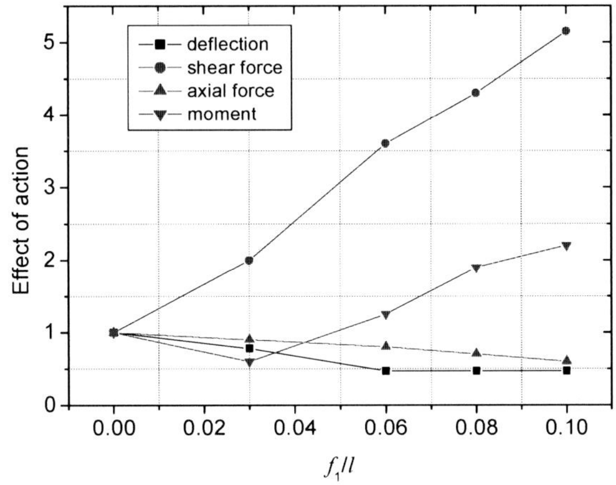

Figure 3: Effect of action versus f1/l curve.

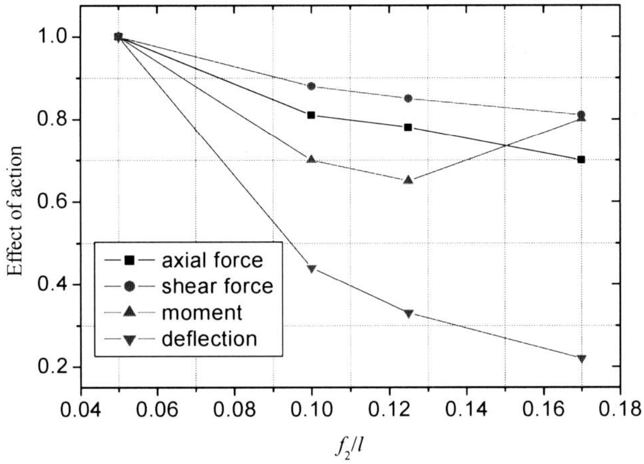

Figure 4: Effect of action versus f2/l curve.

In static analysis, the influence of different factors are taken into consideration, such as high-to-span ratio, tag-to-span ratio, degree of prestress in tendon, and so on. The diagrams of deflection and internal force induced by the uniform distributed load which is applied along the half-span are shown in Figures 3, 4 and 5.

Figure 3 shows that 0.06 is a reasonable high-to-span ratio because the structure provides high rigidity and less internal force. Figure 4 shows that 0.125 is a reasonable tag-to-span ratio because of the same reason. Figure 5 shows that the maximum deflection increases first and then decreases as prestress increases, while moment varies inversely. The maximum axial force and shear force increase, and it is found that these curves vary slowly.

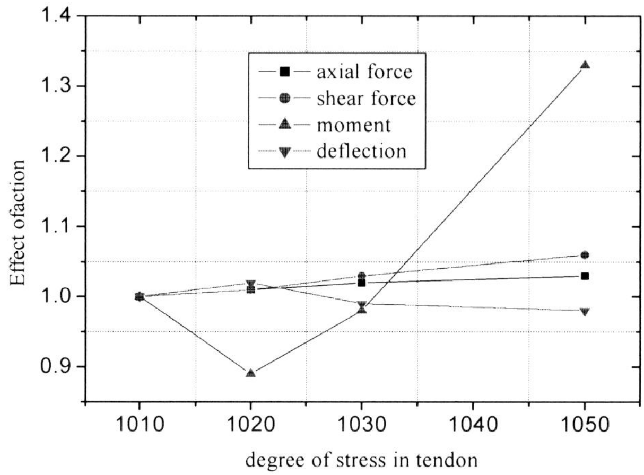

Figure 5: Effect of action versus stress in tendon curve.

4 Dynamic properties of RPC-BSS

With the software ANSYS, the modal analysis of RPC-BSS is carried out and the main vibration frequencies are developed, as shown in Figure 6.

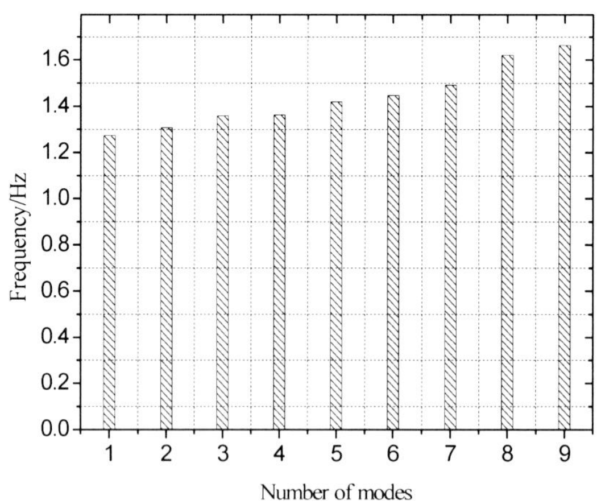

Figure 6: Natural frequency components.

Figure 6 shows that the natural frequencies are densely distributed. And the curve varies slowly.

5 Conclusions

0.06 is a reasonable high-to-span ratio and 0.125 is a reasonable tag-to-span ratio because of the point at which the structure provides high rigidity and induces less internal force. The natural frequencies are densely distributed. RPC-BSS is a type of practicable structure from the point of view of safety, durability, economy, fabrication etc. There are also some problems that are needs to be solved in the future about RPC-BSS, for example, reasonable cross-section, detail design, earthquake resistant behaviour.

References

[1] BAI Z.X. and LIU X.L., A new kind of space structure-beam string structure. Space Structure, 7(2), pp.33-38, 2001.

[2] GAO R. and Jiang L.C., Experimental report on RPC beam structure, 2008, Beijing Jiaotong University, Beijing, P.R. China.

Optimization of concrete/steel thickness ratio in composite steel plate shear walls (CSSWs)

F. Hatami[1], A. Rahai[2] & L. Hoseinzadeh[3]

[1]*Research Institute of Petroleum Industry (R.I.P.I.) & Amirkabir University of Technology, Iran*

[2]*Civil Engineering Department, Amirkabir University of Technology, Iran*

[3]*Civil Engineer, Iran*

Abstract

Composite shear walls made of a thin steel plate with reinforced a concrete cover, which is attached to one or both sides of the steel plate, are the third generation of lateral load resisting shear walls. In addition to increasing resistance, ductility and energy absorption, these walls are very economical and are used in constructing tall buildings, resistance of buildings and tanks. In the present research the effect of various thicknesses of steel plate on the behavior of composite shear walls in different shear connectors has been studied. For this purpose, several different models, including moment frames, steel shear walls and composite shear walls, have been analyzed through numerical and experimental models. In composite shear walls, the number of layers of reinforced concrete cover, the distance between the shear connectors and the thickness ratio of steel plate to the thickness of concrete cover are considered as variables. Increasing the thickness ratio of the steel plate to the concrete layer up to optimum leads to reduction in the out-of plane deflection rate of the steel plate and has no effect on the performance of the composite shear wall. In addition, using two layers of concrete cover on both sides of the steel plate decreases the secondary moment.

Keywords: steel shear wall, composite shear wall, post-buckling, shear connector, concrete cover, steel plate thickness.

WIT Transactions on The Built Environment, Vol 106, © 2009 WIT Press
www.witpress.com, ISSN 1743-3509 (on-line)
doi:10.2495/OP090161

1 Introduction

Using shear walls in different forms of reinforced concrete, steel or the combination of both as a lateral loads resisting system is the common way in structures. Reinforced concrete shear walls have been used for about 30 years, but during the past three decades a comprehensive study on steel shear walls was carried out leading to the application of this innovative system in both new structures and the reinforcement of existing structures. A shear wall with central steel plate core is similar to the vertical plate girder that is used as a lateral load resisting system in medium and high-rise buildings. The system includes separated panels with every panel internally containing two boundary beams and columns. Resistance was also created by the diagonal tension field occurring at the steel plate in combination with the flexural action of the frame. The provision in the Canadian code of design for structural steel buildings confirmed this system and provided the necessary guidance for analysis and design.

Although this modern system has the same capabilities as the other lateral load resisting systems, such as moment frames, concentrically braced frames and eccentrically braced frames, it is better than other systems in the amount of steel used and it is also easier and faster to construct, as well as having a cleaner finished level. The researches shows that this system was more economical in about 50% of cases, considering the amount of used steel to the steel moment frames, this ratio is also about 15% for concentrically braced frames. Compared with reinforced shear walls, these systems have lower weight thus decreasing the earthquake force and consequently foundation dimensions.

Steel shear walls are stiffer compared to different bracing systems, such as X-shapes, and also with respect to the possibility of making openings in each point of the wall it is much more efficient than all other bracing systems. Furthermore, the plastic behavior of this system and the amount of energy absorption is better than other bracing systems. In this system the stress distribution was smoothed out due to greater incorporation, especially in comparison with the bracing system, which causes the suitable homogeneous behavior at the plastic ambient. One of the problems related to the system is out-of plane buckling steel plate, which causes diagonal lines on the steel plate. A smoother distribution of the lines increases the system shear wall capacity. It is possible to use this capability in two ways:

- Using steel grid as a stiffener which is connected to the steel plate, (SSW).
- Using pre-cast concrete or cast I place, which is connected to the steel plate by shear connectors, (CSSW).

Composite shear wall, (CSSW), consisting of a steel plate, is usually used with a concrete cover on both sides. Of course, the cover is located on one side of the steel plate and connects the concrete to the steel plate with shear connectors, such as studs and so on. Composite shear walls are also classified into two types of composite shear walls without gaps and composite shear walls with gaps. Using this system increases the strength and ductility and also decreases the building weight. It also has perfect behavior under seismic loads. In both steel and composite shear wall systems, there is different post-buckling phenomenon. In steel shear walls buckling is generic and applies from resistance of the limited area of the steel plate, while in composite shear walls buckling is converted from generic to the local condition so that the whole steel plate resistance is used.

2 Previous studies

During the past twenty years, a substantial percentage of studies in North America and Japan concentrated on the behavior of these walls under their cyclic loading and analytical models. However, there are valuable results of researches about shear walls available to access and many major constructions were built based on the information from this system about the limitations of seismic designing at current seismic code.

2.1 Research in Canada

Researchers at the University of Alberta (Kulak [4] and Driver et al. [1–3]) have conducted monotonic and cyclic tests of un-stiffened steel plate shear walls. The test results indicate a ductile behavior and significant over-strength. Based on their analytical research, Thorburn et al. [5] proposed an equation for the angle of inclination of the tension field. The test indicated that the proposed equation is sufficiently accurate. Driver et al. [1–3] reported the results of cyclic testing of a four-story steel shear wall specimen. The specimen was a 1/2-scale representation of a dual system with a steel plate shear wall welded to a special moment-resisting frame. The cyclical response of the first floor steel shear wall panel indicates failure mode to local buckling of the column that had occurred during cycle 20, causing large deformation amplitudes at locally buckled areas of the column flange. Researchers at the University of British Columbia have completed a series of cyclic and shaking table tests of steel plate shear walls (Lubell et al. [6], Rezai et al. [7, 8]). In these studies, cyclic shear loads were applied to two single story specimens. The boundary frames in the specimens were moment frames resulting in a "dual" structural system. The two specimens differed only in the base gusset plate details and the top beam. For the second specimen, a stronger base connection and top beam were used. The single story specimens experienced significant inelastic deformations up to ductility of about six. The over-strength was about 1.5. The researchers concluded that the two one story specimens demonstrated that the infill steel plates significantly reduced demand on the moment-resisting frame by producing redundant diagonal story braces that alleviated the rotation demand on the beam-to-column connections.

2.2 Research in Japan

Takanashi et al. [9] and Mimura and Akiyama [10] have conducted some of the earliest tests of steel shear walls. Takanashi et al. conducted cyclic tests of 12 one-story and two 2-story specimens. The 12 one-story specimens had 2.1m width and 0.9m height. They used steel plates with about 2.3mm, 3.2mm and

4.5mm thickness. Compared to typical building dimensions, the specimens could be considered to be 1/4-scale of prototype walls. With the exception of one specimen, all specimens had vertical or vertical/horizontal stiffeners welded on one or both sides of steel plate. The boundary frames were very stiff pin-connected frames. The specimens were loaded along their diagonals to create almost pure shear in the panels. The behavior of specimens was very ductile and drift angels in some cases exceeded 0.10 radians. The shear strengths of the specimens were well predicted by von Misses yield criterion given for pure shear as $V_y=A(F_y/\sqrt{3})$.

2.3 Research in the United Kingdom

Sabouri-Ghomi and Roberts [17] and Roberts [19] have reported results of 16 tests of steel shear panels diagonally loaded. The specimens in these tests consisted of steel plates placed within a 4-hinged frame and connected to it using bolts. Some panels had perforations. The tests indicated that all panels possessed adequate ductility and sustained four large inelastic cycles. Typical hysteresis loops presented in Sabouri-Ghomi and Roberts [17] shows specimens reaching a ductility of more than seven without any decrease in strength. The researchers concluded that the strength and stiffness linearly decreases with the increase in the perforation dimensions.

2.4 Research in the United States

Astaneh-Asl et al. [25] also separated evaluations on the behavior of steel and composite shear walls. In these studies, cyclic loads were applied to 3-story specimens. The specimens were 1/10 scale. The goal was to determine suggestions of seismic designing in the field [24, 25].

2.5 Research in Iran

Alinia et al (2004-2006) numerically evaluated on behavior thin steel shear walls. The results indicated that twisting stiffness of boundary members as ambient frame has great effect on increasing buckling elastic load, but this twisting stiff on post-buckling strength of steel plate was not effective. And increasing steel plate stiffness with stiffener or increasing thickness of plate has low effect on increasing the final strength of steel [26–28].

The results of investigations carried out by Hatami and Sabouri (2005) indicates that the kind of beam to column connection has no critical effect on behavior of plates [29].

Many different experiments were carried out by Hatami and Rahai (2008) on behavior of composite shear walls made of concrete and steel or concrete and carbon polymer fibers in research Center of Structure and Earthquake of Amirkabir Industrial University that their results are about to be published [30].

3 Description of different parts of the composite shear wall system for its performance

Steel plate is a main element in composite shear wall. Preferably, this plate is selected of steel with low yield point. For example St37 steel plate is preferred for high strength steel plate. Because St37 steel plate due to its low yield point is preferred to encourage yielding of steel plate.

The reinforced concrete cover on one side or both sides of a steel plate carries some of the story shear by developing compression diagonal field and increased strength and stiffens. Of course main role of the reinforced concrete cover is to prevent out-of plan buckling of steel plate prior to reaching yielding. Shear connectors are used to concrete is to prevent buckling of steel plate. In some case shear studs not only are subjected to shear but also to a considerable tension due to local buckling of the steel plate. For cast-in-place concrete, welded shear studs are usually used, for pre-cast concrete walls bolts can be used.

4 Models properties and investigated parameters

In table 1 the properties of constructed models presented for thickness of concrete cover and shear connectors arrangement and thickness of steel plate.

Table 1: Denomination of models (rate per mm).

No	Model	Thickness of Concrete Layer	Studs Number	Thickness of Steel Plate
1	MRF	-	-	-
2	SSW	-	-	2
3	CSW-5-3*3 (3)	50	9	2
4	CSW-5-3*8 (3)	50	24	2
5	CSW-5-3*3 (5)	50	9	5
6	CSW-5-3*8 (5)	50	24	5
7	CSW-5-3*3 (10)	50	9	10
8	CSW-5-3*8 (10)	50	24	10

5 Geometrical properties of models

The length of frame span that is center distance to center of both sides column is 2000mm and height is also frame 1000-2000mm which is center distance to center of top and bottom beams. Columns and beams sections were also double IPE2000 that were reinforced with two plates (150*12) on the flanges.

Steel plate inscribed inside the boundary beams and columns, it is 1776*776 long and 1776*1776mm wide, and the thickness of these plates are also considered 3, 5 and 10mm. The thickness of concrete cover is 50 and 100mm. there is a 30mm gap between the concrete cover and the boundary elements, 716*1716, 1716*1716mm length and width. In order to connect the concrete cover with 50mm thickness to steel plate using 7mm diameter and 100mm length bolts and for connecting the concrete cover with 100mm thickness using bolt with 10mm diameter and 120mm length. To reinforce concrete cover thickness 50mm using 6.5mm diameter bars with center distance to center 65mm and for concrete cover 100mm using 8mm diameter bars with center distance to center 85mm.

6 Mechanical properties of material

Characteristics of steel that is used in all models St37 with yield point 240MPa and ultimate stress 370MPa and the behavior of steel bilinear elastoplastic curve is applied for modeling. The concrete has also compression strength of 28 day's cylindrical core sample 45Mpa and its tension strength is equivalent to 3MPa.

7 Modeling theories

To analyze model, the following theories were considered:
- Specimens at their lower level fixed on all directions.
- The behavior of used steel such as steel plate, beams, columns and bars inside the concrete considered bilinear.

Meshing of models was also selected so that connecting the flange to web at columns and beams profile and steel plate to the boundary elements in specimens were rigid connections. Mesh dimensions was also maximum 50*50mm. Using elements on the study, includes Solid 65 for modeling reinforced concrete and considering its crack and Shell 143 for steel plate and beams and columns and Beam188 for shear connectors.

8 Loading

Applied load on models:

A. Static loading: 1. Static loading up to 600KN & 2. Static loading so that it leads to destroy the model

B. Harmonic loading-frequency 1/60HZ and amplitude value 600KN

At harmonic loading the cyclic loading was gyred and the rate of frequency of harmonic loading is 1/60HZ and the rate of amplitude value of the force is 600KN that lasted for 540Sec.

9 Comparing the results for the numerical and experimental models

To construct validity of analyzed model of composite shear wall specimen (1000*2000mm) on which the thickness of steel plate and concrete cover is 3 and 50mm respectively, and height overall ratio to the span is 1/2. Applied under harmonic loading and its deformation-load diagram drown and compared to the analyzed results of similar numerical model (Figure 2). This experimental model is shown in Figure 1.

Figure 1: Experimental comparative model.

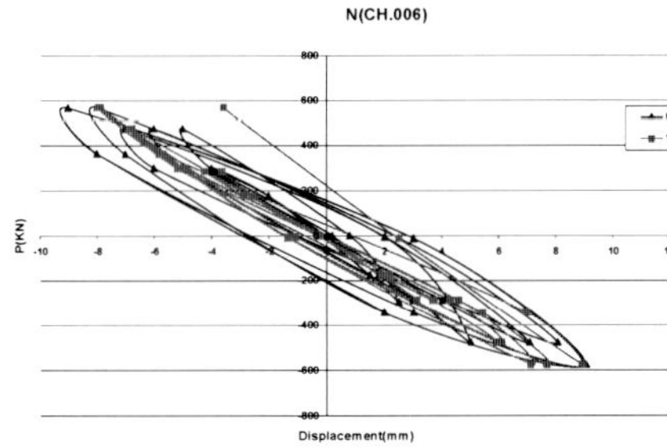

Figure 2: Comparing results of deformation-load in the numerical and experimental model for SSW.

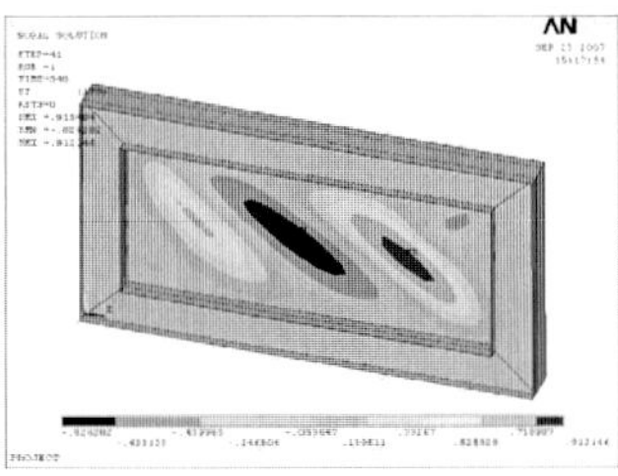

Figure 3: Out-of plane deflection of steel shear wall (SSW).

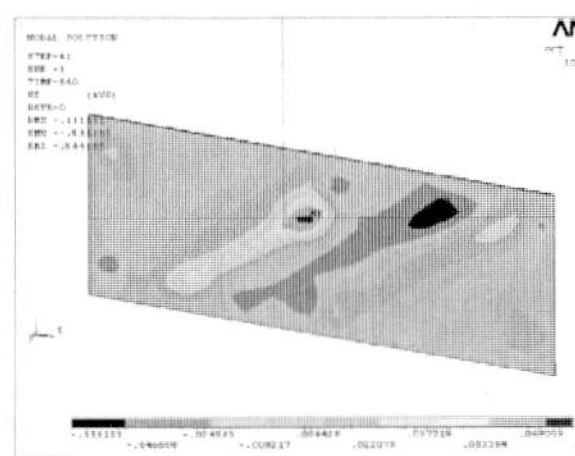

Figure 4: Out-of plane deflection of steel shear wall (CSW-5-3*3).

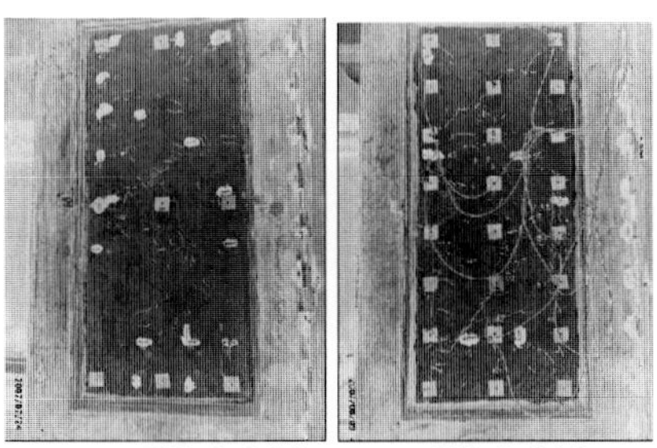

CS-5-3*8 CS-5-3*3

Figure 5: Comparing ultimate cracks at testing specimens.

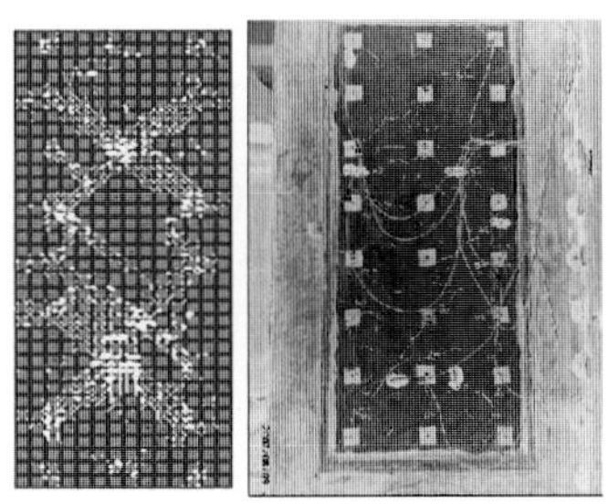

Figure 6: Comparing cracks of experimental with numerical models (CS-5-3*8).

It can be observed that the discrepancy of area under hysteresis curve for numerical and experimental is less than 8%.

10 Software output results in harmonic loading

At the last harmonic loading step, at step 41 where the time of loading is 540Sec after beginning the results of software output and experimental study graphically presented in figure 3–6 to see and compare out-of plane deflection of steel and composite shear wall and also formation of cracks in different experimental and numerical specimens.

11 Comparing the results of numerical models analysis

11.1 Effect of various distances between shear connectors

As is clear from tables 2 and 3, when the number of shear connectors increases (decreasing distance between shear connectors), the rate of lateral displacement decreased 25-30% and the rate of stiffness increased 30-35%.

11.2 Effect of various concrete cover to steel plate thickness ratios

In figure 7, the effect of various concrete to steel thickness ratio (invariable concrete thickness) is studied. This study includes in-plane and so, out-of plane deflection, steel stress, concrete stress and bolts stress.

As it can be observed in table 4 the increasing thickness of steel plate from 3mm to 5mm, the rate of in-plane and out-of plane displacement decreases so that this reduction is about 35-40%, but when the thickness of steel plate

Table 2: Increasing stiffness ratio and decreasing lateral displacement for models at harmonic loading.

Model	FRAME (MRF)	SSW	CSW-5-3-3	CSW-5-3-8
Increasing of Stiffness ratio	1	2.43	3.03	3.7
Decreasing of Lateral Displacement ratio	1	0.41	0.33	0.27

Table 3: Various ratio out-of plane displacements for models at harmonic loading.

Model	SSW	CSW-5-3-3	CSW-5-3-8
Decreasing of out of Plane Displacement ratio	1	0.63	0.37

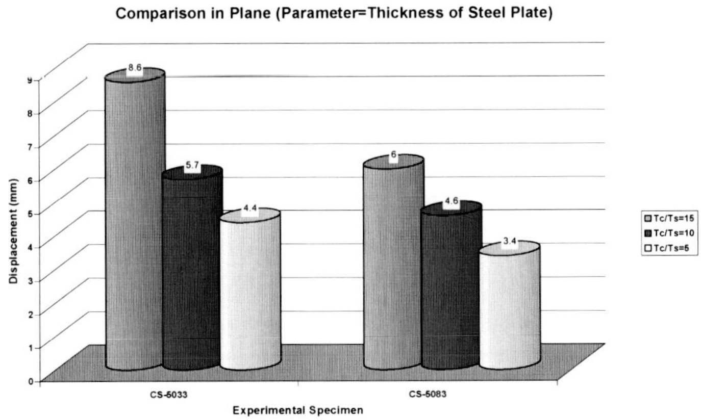

Figure 7: In-plane deflection for various thicknesses of steel plate.

Table 4: Various ratios of case study parameters and the effect of various concrete cover to steel plate thickness ratios in model 50-3-3.

Model	Max. Stress (Steel)	Max. Stress (Concrete)	Out of Plane Disp.	In Plane Lateral Disp.
50-3*3 (3)	1	1	1	1
50-3*3 (5)	0.59	0.89	0.6	0.66
50-3*3 (10)	0.42	0.68	0.49	0.51

Table 5: Various ratios of decrease in case study parameters for using two concrete layers on the sides.

Response / Model	In Plane Lateral Disp.	Out of Plane Disp.	Max. Stress (Steel Plate)	Max. Stress (Con. Layer)	Max. Stress (Shear Studs)
50-3*8	1	1	1	1	1
2*50-3*8	0.96	0.78	0.97	0.93	0.97
100-3*8	0.83	0.68	0.95	0.67	0.88

increases from 5mm to 10mm, there is no substantial in-plane decreasing and out-of plane displacement was about 10-15%. Therefore it is concluded that to minimize a specific out-of plane value of thickness displacement and increment of the thickness has no effect on decreasing displacement.

As is shown in table 5 the rate of in-plane lateral displacement to the one concrete layer on one side of steel plate decreased just 4% due to use of two concrete layer on both sides. This is not substantial. So, the rate of various out-of plane displacements decreased about 22%. Hence, when there is just one concrete layer on one side of steel plate, P-Δ (secondary moment) considered and add with the rate of out-of plane displacement caused by buckling phenomenon while effect of P-Δ decreased using two concrete layer on both sides of steel plate and the rate of out-of plane displacement is due to plate buckling. Existence of a concrete layer on one side of steel plate having thickness equivalent twice thickness of every concrete on both sides decreased the rate of displacement and stresses.

12 Summary and conclusion

1. In case of using double concrete layers on both sides of steel plate, there are no substantial variations in the rate of in-plane lateral displacement. The composite shear wall has more capability in energy absorbing, extending the stress on more points of steel plate and ductility to the steel shear wall.
2. Increasing thickness ratio of reinforced concrete cover to the steel plate, increased stiffness and ductility of composite shear wall. Stiffness and ductility increased in field of panel size and geometry and further more it has no substantial effect.
3. Various distance between shear connectors, cause various ductility and energy absorption in composite shear wall. (for example by changing the distance of shear connectors between 440 to 195mm (decrease 56%) stress distribution between shear connectors when it is doubled the stress of steel plate will decrease about 21% and the stress of concrete plate will increase about 33%. That indicate transformation a part of steel plate stress to the reinforced concrete layer increased and cause the increasing the rate of energy absorbing of composite shear wall.
4. Variations in the thickness of reinforced concrete cover to the steel plate at composite shear wall up to a specific value can increase ductility, stiffness and energy absorbing. With respect to the analysis, this ratio obtained for composite shear wall as 1:1 and 1:3 scales equivalent 7.74 and 16.4 respectively, with respect to the thickness of 7mm and 3.5mm of steel plate, in both cases the optimum thickness of reinforced concrete cover determined about 50mm at monotonic and cyclic loading.
5. Distance between shear connectors on composite shear wall up to a specific value can improve system behavior

References

[1] Driver RG, Kulak GL, Kennedy DJL, Elwi AE. Cyclic test of four-story steel plate shear wall. J Struct Eng ASCE 1997; 124(2):112-20.

[2] Driver RG, Kulak GLDJ, Elwi AE, Kennedy L. FE and simplified models of steel plate shear wall. J Struct Eng ASCE 1997;124(2): 121-30.

[3] Driver RG, Kulak GL, Elwi AE, Kennedy DJL. FE and simplified models of steel plate shear wall. Journal of Structural Engineering ASCE 1998; 124(2):121-30.

[4] Kulak GL. Unstiffened steel plate shear walls. In: Narayanan R, Roberts TM, editors. Structures subjected to repeated loading. New York: Elsevier; 1991. p. 237-76 [chapter 9].

[5] Thorburn L.J.; Kulak G.L.; Montgomery C.J.; "Analysis of steel plate shear walls"; Structural Engineering report No. 107; 1983; University of Alberta; Canada.

[6] Lubell AS, Prion HGL, Ventura CE, Rezai M. Unstiffened steel plate shear wall performance under cyclic loading. J Struct Eng ASCE 2000; 16(4):453-60.

[7] Rezai M.; Ventura C.E.; Prion H.G.L.; "Numerical investigation of thin unstiffened steel plate shear walls"; Proc. 12th World Conference on Earthquake Engineering; 2000.

[8] Rezai M.; Ventura C.; Prion H.; "Simplified and detailed finite element models of steel plate shear walls"; Proc. 13th world conference on earthquake engineering; Vancouver; B.C.; Canada.

[9] Takanashi Y.; Takemoto T.; Tagaki M.; "Experimental Study on Thin Steel Shear Walls and Particular Bracing under Alternative Horizontal Load"; IABSE, Lisbon, Portugal.; 1973.

[10] Mimura H.; Akiyama H.; " Load-Deflection Relationship of Earthquake – Resistant Steel Shear Walls with a developed Diagonal Tension Field"; October; 1977.

[11] Sugii K.; Yamada M.; " Steel Panel Shear Walls with and Without Concrete Covering"; 11th World Conference on Earthquake Engineering; Acapulco; Mexico; 1996.

[12] Torii S.; Teramoto T.; Kihara H.; Kitamura H.; "The Response Control Design of High-rise Building with Low Yield Steel Wall"; Proceedings on CD-Rom; 11th World Conference on Earthquake Engineering; Acapulco; Mexico; 1996.

[13] Chen SJ, Yen CZ. Application of L YP steel shear damper for seismic resistance. Struct Eng 2000; 15(1):3-21.

[14] Chen SJ, Yeh CH. Enhancement of ductility of steel beam-to-column connections for seismic resistance. Proceedings of the SSRC 1994 technical session, Structural Stability Research Council. Lehigh University, PA.

[15] Chen SJ, Yeh CH, Chu JM. Ductile steel beam-to-column connections for seismic resistance. J Struct Eng ASCE 1996; 122(11):1292-9.

[16] Chen Sheng- Jin, Jhang Chyuan. Cyclic behavior of low yield point steel shear walls. Thin-Walled Structures 44(2006) 730-738.

[17] Sabouri-Ghomi S, Roberts TM. Nonlinear dynamic analysis of steel plate shear walls including shear and bending deformations. Engineering Structures 1992;14(5):309-17.

[18] Sabouri-Ghomi S, Ventura CE, Kharrazi MHK. Shear analysis and design of ductile steel plate walls. J Struct Eng ASCE 2005; 131(6): 878-89.

[19] Roberts T.M.; "Seismic Resistance of Steel Plate Shear Walls"; Engineering Structures 17 (1995), 344-351.

[20] Caccese V, Elgaaly M, Chen R. Experimental study of thin steel-plate shear walls under cyclic load. J Struct Eng ASCE 1991; 119(2): 573-88.

[21] Elgaaly M, Caccse V, Du C. Postbuckling behavior of steel-plate shear walls under cyclic loads. J Struct Eng ASCE 1993; 119(2): 588-605.
[22] Elgaaly M, Liu Y. Analysis of thin steel plate shear walls. Journal of Structural Engineering ASCE 1997; 123(11):1487-96.
[23] Elgaaly M. Thin steel plate shear walls behavior and analysis. Thin-Walled Structures 1998; 32:151-80.
[24] Zhao Q, Astaneh-Asl A. Cyclic behavior of traditional and innovative composite shear walls. J Struct Eng ASCE 2004.
[25] Astaneh-Asl A. Seismic behavior and design of steel shear walls. Steel TIPS report. Moraga (CA): Structural steel educational council; 2001.
[26] Alinia MM. A study into optimization of stiffeners in plates subjected to in-plane shear loads. Thin-Walled Structures 2005; 43(4):845-60.
[27] Alinia MM, Dastfan M. Effect of surrounding members on shear buckling of panels. Thin-Walled Structures 2004
[28] Alinia MM, Dastfan M. Behavior of thin steel plate shear walls regarding frame members. Journal of Constructional Steel Research 62 (2006), 730-738.
[29] Hatami, F. & Sabouri, S., Behavior of Steel Plate Shear Walls in Earthquake Due to Change of Rigidity of the Internal Storey Beams. Amirkabir Journal, Vol. 15, No 60-2, Civil Engineering, Tehran, IRAN, Fall & Winter 2004-2005.
[30] Hatami, F, Rahai, A. Performance Evaluation and Optimization of Composite Shear Wall. Ph.D. Dissertation, Amirkabir University of Technology, January 2008.

Section 5
Design optimisation in aircraft engineering

Design space investigation by Response Surface Model techniques in aeronautical metal cutting applications

A. Del Prete[1], A. A. De Vitis[1] & D. Mazzotta[2]
[1]*Department of Engineering Innovation, University of Salento, Italy*
[2]*AVIO S.p.A., Italy*

Abstract

A finite element model (FEM), which is able to predict the cutting interactions between the tool and the workpiece (cutting forces and cutting edge temperature) for a typical aeronautical engine Nickel-alloy Inconel 718, has been used to perform a reduced number of FEM calculations indicated by a DOE, in order to evaluate the design space for process parameters. These well-distributed results can be subsequently used to create and evaluate the quality in terms of correct response behaviour for the metamodels created through different approximation techniques (polynomial and neural network). The metamodels based on the best methodology (in terms of effectiveness of process behaviour prediction) have been used to optimize, through Adaptive Simulated Annealing (ASA) algorithms, the process parameters defined in a CAM part program block. The aim of the authors is to modify the Part Program operation parameters according to the constraints arising from the physical nature of the cutting process obtained by FEA.
Keywords: machining, FEM, optimization, RSM, ASA.

1 Introduction

The nickel-alloy Inconel 718 is usually well-used in the aerospace industry for the manufacture of components that require: lighter, harder, stronger, tougher, stiffer, more corrosion- and erosion-resistant materials capable of maintaining their mechanical properties at elevated temperatures, such as in jet engines. The unique and desirable heat-resistant characteristics of superalloys, on the other hand, impair their ability to be worked due to the extremely high temperature

WIT Transactions on The Built Environment, Vol 106, © 2009 WIT Press
www.witpress.com, ISSN 1743-3509 (on-line)
doi:10.2495/OP090171

generated at the cutting edge. These tough working conditions tend to deform the cutting tool, leading to accelerated wear during machining, particularly at higher speed conditions. The ability to be worked of aerospace alloys will continue to decline as new materials are developed to meet the increasing demand for higher temperature-resistant materials and for more efficient aero-engines (Ezugwu [1]). Considerable research and development efforts have been directed, worldwide, towards an improvement of the machining operations in order to ensure an efficient and cost effective machining of these superalloys and to understand their behaviour when they are machined at higher cutting conditions (Malinov *et al.* [2]). A good understanding of the behaviour and of the relationship among the workpiece materials, the cutting tool materials, the cutting conditions and the process parameters is an essential requirement for the optimisation of the cutting process. A significant improvement in process efficiency may be obtained with process parameters optimization that identify and determine areas of critical process control factors. The availability of such a methodology allows engineers to choose the desired outputs or responses with acceptable variations ensuring a lower cost of manufacturing (Montgomery [3]). The selection of optimal machining conditions is a key factor in achieving this condition (Tan and Creese [4]). The first necessary step for process parameters optimization in any metal cutting process is the ability to understand the principles governing the cutting processes by developing an explicit mathematical model. The resulting model provides the basic mathematical input required for the formulation of the process objective function. An optimization technique provides an optimal or near-optimal solution to the overall optimization problem formulated, which is subsequently implemented in the actual metal cutting process. One of the several optimization modelling techniques proposed and implemented is based on response surface design. Many researchers use Response Surface Models (RSM) to optimize problems in metal cutting process parameters. Taramen [5] uses a contour plot technique to simultaneously optimize tool wear, surface finish, and tool force for finished turning operation. Lee *at al.* [6] provide an interactive algorithm using both RSM and mathematical modelling to solve a parameters optimization problem in the turning operation.

2 Metamodels construction approach

The RSM is a dynamic and very important tool of design of experiment (DOE), in which the relationship between the responses of a process with their input decision variables are mapped to achieve the objective of maximization or minimization of the response properties. The first necessary step in RSM is to map responses, i.e. Y as a function of independent decision variables (X1,.,Xn). Normally RSM techniques are based on a series of experimentation, and may not be feasible or cost effective for manufacturers in many manufacturing situations. In this study, for the RSM construction the authors have used data drawn from the numerical simulations. In this way, if the simulation code has been previously validated through experimental-numerical correlation, it becomes possible to avoid costly experimental campaigns for data acquisition.

In order to model the true functions through metamodels based on data obtained with FEA, the authors have performed 40 simulations (with a version of Latin Hypercube DOE, called *Optimal Latin Hypercube*) based on design points selected within the two-dimensional design space, which has the following boundaries: feed for tooth (f) [0.04; 0.25 mm/rev]; cutting velocity (v_t) [100; 230 mm/sec.]. In this DOE technique an optimization process is then applied to this initial random Latin Hypercube design matrix. By swapping the order of two factor levels in a column of the matrix, a new matrix is generated and the new overall spacing of points is evaluated. This optimized process designs a matrix in which the points are spread as evenly as possible within the design space defined by the lower and upper level, *iSIGHT user's manual* [7].

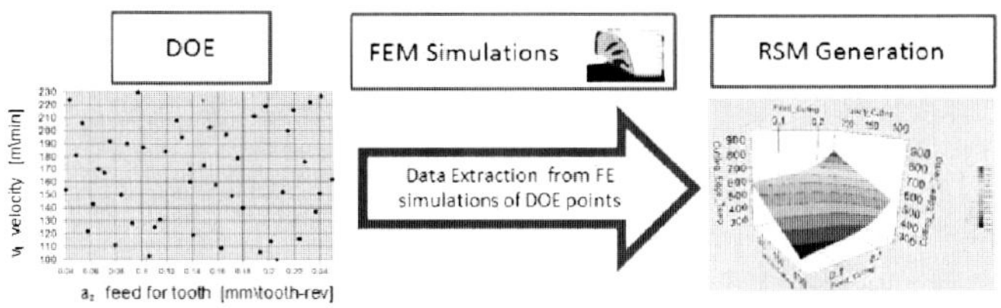

Figure 1: Metamodels construction procedure.

3 Theory of metamodels based on polynomial approximations

The RSMs used in this work are based on polynomials approximation used in models of the second, third and fourth order. The models based on the approximation of the third and fourth order do not have any mixed polynomials for terms of the upper second order (see table 1 the comparison), but only pure cubic and quartic terms are included to reduce the amount of data required for the model construction (eqn. (1)).

$$\overset{\approx}{F}(x) = a_0 + \sum_{i=1}^{N} b_i x_i + \sum_{i=1}^{N} c_{ii} x_i^2 \sum_{ij(i<j)}^{N} c_{ij} x_i x_j + \sum_{i=1}^{N} d_i x_i^3 + \sum_{i=1}^{N} e_i x_i^4 \qquad (1)$$

where: N is the number of model inputs; x_i is the set of model inputs; a, b, c, d, are the polynomial coefficients (*iSIGHT user's manual* [7]).

Table 1: Number of coefficients in polynomials involving two input parameters.

	Number of Coefficients	
Polynomial Degree	Complete Formulation	Reduced Formulation
2	6	6
3	10	8
4	15	10

4 Theory of metamodel based on artificial neural network

Radial Basis Functions (RBF) are a type of neural network used to approximate the behaviour of several types of systems. They adopt a hidden layer of radial units and an output layer of linear units and they are characterized by reasonably fast training and reasonably compact networks. Weissinger [8] was the first to use radial basis functions to calculate the flow around wings. This neural network utilizes the Gaussian curve to map values. The network has n inputs and k outputs. The radial basis network is a very efficient network when function approximation is needed, because it has the ability to represent nonlinear functions.

5 Simplified finite element 2D model and data extraction methodology

The analyzed operation is a down cut mill operation. In the climb mode shown in fig. 2(A), the feed on each tooth is bigger at the point of initial contact with the workpiece and becomes very small at the end of the tooth-workpiece engagement.

The maximum force exchanged between the tool and the workpiece occurs when the tooth surface impacts on the workpiece. To detect components F_x and F_y, a simplified 2D model has been chosen (fig. 2(B)), where the real feed per tooth is the DOC on the finite element model ($a_{z,\ max}$) and the cutting speed is equal to the tangential velocity of the cutter (V_t). The values of the exchanged forces between the tool and the workpiece used in the study were detected when the value of the observed forces in the simulation post processing can be considered stable. Cutting temperature data utilized to create metamodels are an average of the temperature extracted in five fixed nodes detected on the cutting edge and meshed when the thermal steady state condition on the tool-workpiece interface was reached (after about 2 mm of cutting length).

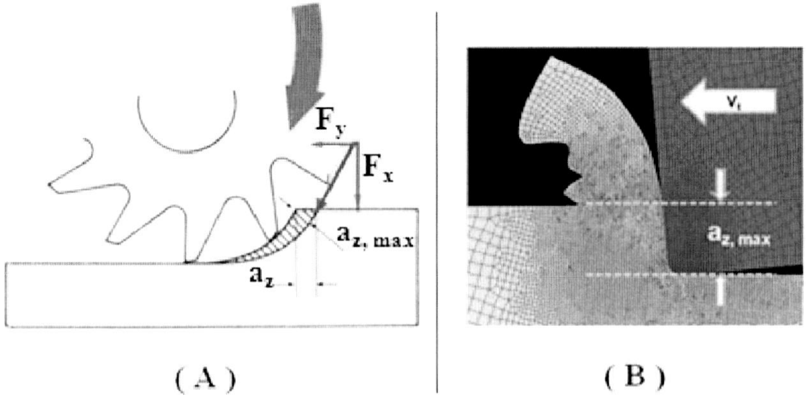

Figure 2: (A) Climb mill operation; (B) FE simplified 2D model.

6 Finite element model

Finite element (FE) analysis of the dry milling operations with the circular insert in WC and no wear have been carried out. A two-dimensional plane-strain thermo-mechanical analysis, based on the updated Lagrangian formulation, was performed using the implicit FE commercial code properly developed for machining FEA. For this purpose, the workpiece (dim 4x1.5 mm) was initially modelled with 6500 bilinear four-node quadrilateral elements, with dimensions respectively of 0.007 mm along the cutting edge, 0.01 mm on the first 0.3 mm on machined surface and 0.15 mm in the remaining area. The insert in tungsten carbide (WC) with 8% in Cobalt has been characterized using the software default material library. The tool orthogonal rake angle was γ= -7°, the cutting edge radius was r_ε=0.05 mm and it has been modelled as rigid. It has been meshed and subdivided into 1500 elements with the very small dimension of 0.004 mm, in the nose zone in order not to lose the small radius of curvature in this area. The material properties of the cutting insert have been characterized using the software default material library. A constant frictional stress law on the rake face is assumed equal to a fixed percentage of the shear flow stress of the machined material (eqn. (2)).

$$\tau = m \cdot k \qquad (2)$$

where k is the shear flow-stress of the workpiece material and m is a friction factor, assumed to be equal to 0.5. A value of 100 kW/m^{2}K has been adopted for the interface heat transfer coefficient, h. For the workpiece made of Inconel 718, proper values were defined for the following characteristics: Young's module; Poisson's ratio; thermal expansion; heat capacity; emissivity; thermal conductivity. Corresponding values are experimentally defined in the range of temperature [20°C; 1200°C] by the industrial partner AVIO S.p.A. For workpiece material characterization in the plastic field a Johnson Cook constitutive model it has been adopted, see eqn. (3) (Bil *et al.* [9]).

$$\sigma_{eq} = \left(A + B\varepsilon^n\right)\left[1 + C\ln\left(\frac{\dot{\varepsilon}}{\dot{\varepsilon}_0}\right)\right]\left[1 - \left(\frac{T - T_{room}}{T_{melt} - T_{Room}}\right)^m\right] \qquad (3)$$

where ε is the plastic strain, $\dot{\varepsilon}$ is the strain rate (s^{-1}), $\dot{\varepsilon}_0$ is the reference plastic strain rate (s^{-1}), T is the temperature of the work material (°C), T_{melt} is the melting temperature of the work material (1400 °C) and T_{room} is the room temperature (20 °C). Coefficient A is the yield strength (MPa), B is the hardening modulus (MPa), C is the strain rate sensitivity coefficient, n is the hardening coefficient and m is the thermal softening coefficient. For the constants present in the J&C equation the authors have used the follow values: ε_0=1.001 [1/sec.], A = 450 [MPa], B = 1700 [MPa], C = 0.017, n = 0.65, m = 1.3 (Uhlmann *et al.* [10]).

7 Quality metamodels evaluation

The approximation error analysis provides a visual representation of the quality of an approximation model for each response. The total error is calculated for each response using the average error. The differences between the actual (workflow execution) and the predicted (approximation model execution) values for all error samples are averaged and then normalized by the range of the actual values for each response. Normalizing the error value allows one to compare the error level of different responses with different magnitudes in respect to approximation model predictions. The error is calculated based on a number of sample points (19 in this case) specifically allocated for error analysis. From the average approximation error analysis (table 3) it is possible to evaluate how the metamodels that give the best approximation of the real behaviour of the three analyzed responses are the ones created with the approximation technique based on the neural networks (RFB). The resulting data confirm that for the metamodels based on the polynomial approximation, the confidence of prediction of the analyzed responses behaviour is much closer to reality as is the larger grade of polynomial used to create the metamodel itself.

Table 2: Comparison of % average error for analyzed responses for each examined approximation technique.

Response	2° Order	3° Order	4° Order	RBF
T_{max}	12.2%	11.6%	9.7%	0.9%
F_x	1.1%	1.0%	0.6%	0.04%
F_y	3.9%	3.8%	3.0%	0.1%

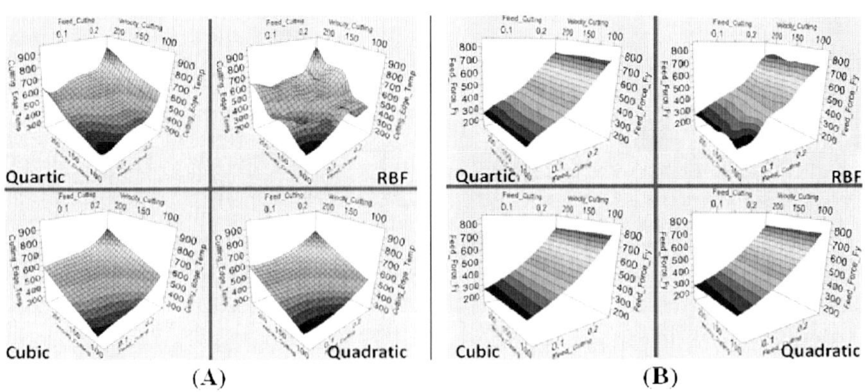

Figure 3: Metamodel shapes comparison for the four adopted techniques. (A) for T_{max} response; (B) for F_y response.

In fig. 3(A) for the response cutting edge temperature, it is possible to appreciate how the shape of the response surface varies a lot, ranging from the one created starting with second grade polynomials (quasi-linear) to the one created with the RBF technique that presents a much less regular surface where there are local maxima and minima. The metamodels of the thrust force F_y (fig. 3(B)), also show a more linear performance in the RS of the fourth grade or in the RBF. Similar behaviour has been found for the cutting force F_x. The highly non-regular behaviour of the cutting temperature has been highlighted by the detected average error. The latter is significant in the metamodels created with a low polynomial grade and cannot predict the real values for highly non-linear responses.

8 Example of CAE-CAM optimization procedure based on metamodels

Metamodels created with the most efficient methodology in predicting the three responses behaviour, namely the response surfaces created with RBF, constitute the basis for the optimization phase. This procedure has a considerable advantage; in the optimal solution research path, the optimization algorithm performs the iteration run using metamodels and it is for this reason preferable to the long calculation time typical of the calculation run of FEA (fig. 4).

In this study the authors have used, in order to optimize the process parameters, a modified version of the *Simulated Annealing* search algorithm (SA), called *Adaptive Simulated Annealing* (ASA). The distinct feature of this method is the temperature change mechanism, which is an important part of the transition probability equation. In conventional simulated annealing, the search begins with high temperature allowing a higher chance of transition to a worse solution. By doing so, the search is able to move off from local minima. However, as the search continues, the temperature continuously declines resulting in a reduced chance of uphill transition. Such an approach could be useful if the local minima are near the start point, but it might not lead to a near-optimal solution if some local minima are encountered at a relatively low temperature towards the end of the search. To overcome this difficulty, the adaptive simulated annealing method uses an adaptive cooling schedule, based on the profile of the search path, which adjusts the temperature dynamically. Such adjustments could be in any direction, including the possibility of reheating in order to push the search out of the local minima. This optimization activity aims to minimize the execution time through the maximization of the feed rate in compliance to physical micro-scale limits of the cutting process (max cutting temp., cutting forces). Figure 5 shows the optimization scheme adopted.

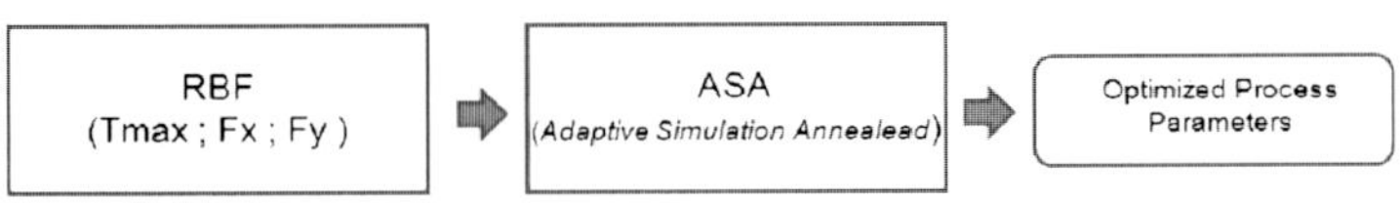

Figure 4: Flowchart of the adopted optimization procedure.

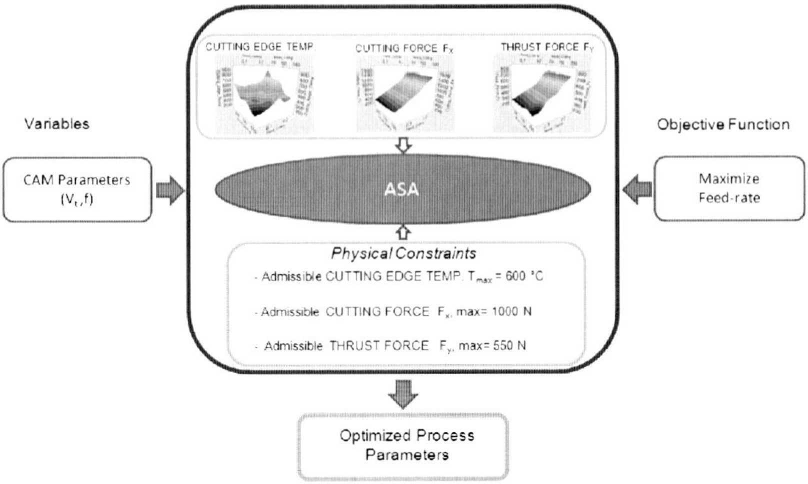

Figure 5: Scheme of optimization problem definition.

The problem can be stated as:

Variables: - $Vt \in [100\ ;\ 230]$ *mm/sec.*

- $f \in [0.04\ ;\ 0.25]$ *mm/tooth - rev*

Constraints: – *Max admissible Cutting Edge Temp.* $T\max = 600^{\circ}\ C$

– *Max admissible Cutting Force* $Fx = 1000\ N$

– *Max admissible Thrust Force* $Fy = 550\ N$

Objective function: – *Maximize feed rate Va mm/min*

The process parameters present in the analyzed block of the part program are feed and speed. They have been taken into consideration with these values: 1) feed (F) = 50 [mm / min]; 2) speed (S) = 96 [rpm]. Their respective values have been converted into the units of measurement used in the 2D finite element model and to define metamodel design space, f = 0.13 mm/tooth-rev and V_t = 209 mm/sec. After the optimization phase the new set of process parameters has been converted into the units of measurement used in the part program ($V_{t,opt}$= 225 mm/sec. to S = 107 rpm and f_{opt} = 0.18 mm/tooth-rev to F = 77 mm/min). Then, they have been incorporated into the analyzed part program block (fig. 6).

The process parameters indicated by the optimized block present an increase in the cutting speed of 11.5% and an increase in the feed rate of 38.4%. In fig. 8, the diagrams show the iterations carried out by the ASA searching the optimal solution.

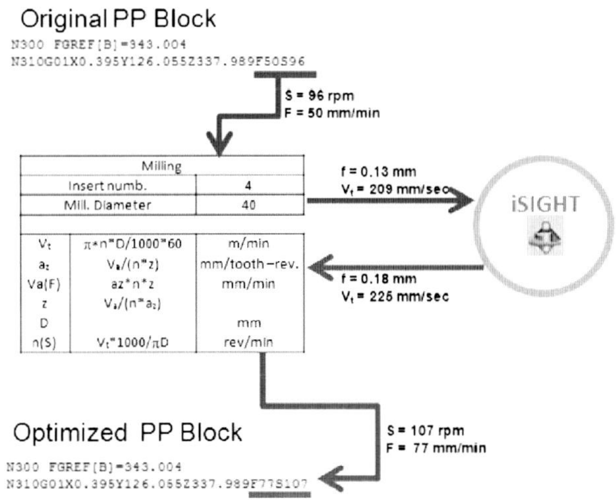

Figure 6: Flowchart of the input-output process parameters optimization.

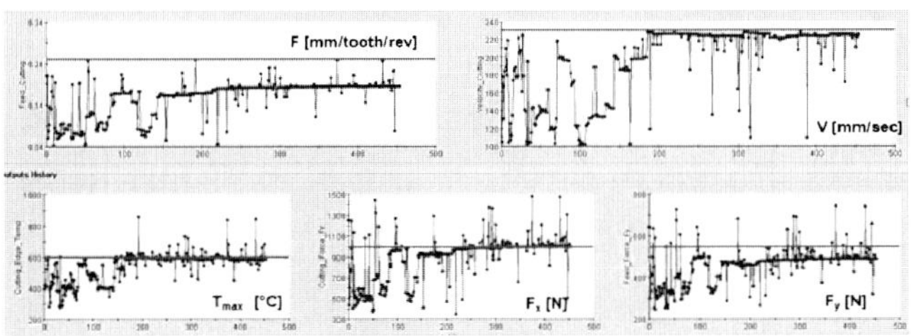

Figure 7: ASA iterations history.

9 Conclusions and further developments

In this article the authors have analysed the potentiality of four methods for the creation of response surfaces in order to thoroughly predict the behaviour of some of the typical responses of the physical phenomena of metal cutting processes.

The response surfaces have been created from data extracted from FEM simulations performed on the basis of a DOE. The approximation method that has proved to be the most effective was the one based on neural networks (RBF). The response surfaces created with this methodology have been later used as a basis for the optimization through an ASA algorithm of the process parameters of a part program block. In the future, the authors will focus on the search for numerical-experimental correlation of the FEM model in order to improve the accuracy of metamodels. Another research area will focus on the adjustment of CAE-CAM integration procedures in order to obtain automated process parameters optimization compliant to the physical limitations of the process and their replacement in the part program that has to be optimized.

References

[1] Ezugwu, E. O., Advances in the machining of nickel and titanium base superalloy. *Japan Society for Precision Engineering Conference*, pp. 1-40, 2004.
[2] Malinov, S. Sha, W. McKeown, J.J., Modelling the correlation between processing parameters and properties in titanium alloys using artificial neural network. *Computational Materials Science*, **21(3)**, pp. 375-394, 2001.
[3] Montgomery, D. C., *Introduction to statistical quality control* (2end ed.), Wiley, New York, pp. 86-87, 1990.
[4] Tan, F. P., & Creese, R. C. A generalized multi-pass machining model for machining parameter selection in turning. *International Journal of Production Research*, **33(5)**, pp. 1467–1487, 1995.
[5] Taramen, K., Multi-machining output-multi independent variable turning research by response surface methodology. *International Journal of Production Research*, **12(2)**, pp. 233–245, 1974.
[6] Lee, Y. H., Shin, H. M., & Yang, B. H., An approach for multi-criteria simulation optimization with application to turning operation. *Computers and Industrial Engineering*, **30(3)**, pp. 375–386, 1996.
[7] Engineous Software - *iSIGHT Version 3.0 User's Guide* - 2008.
[8] Weissinger, J. *Lift distribution of swept-back wings*. NACA pp.1120, 1947.
[9] Bil, H., Kilic, E., Tekkaya, A. E., A comparison of orthogonal cutting data from experiments with three different finite element models. *International journal of Machine Tools & Manufacture*. **44(9),** pp. 933-944, 2004.
[10] Uhlmann, E., Von Der Schulenburg, M.G., Zettier, R., Finite Element Modelling and Cutting Simulation of Inconel 718. *CIRP Annals - Manufacturing Technology,* **56(1),** pp. 61-64, 2007.

Optimum aerofoil parameterization for aerodynamic design

R. W. Derksen[1] & T. Rogalsky[2]
[1]*Department of Mechanical and Manufacturing Engineering, University of Manitoba, Winnipeg, Canada*
[2]*Mathematics Department, Canadian Mennonite University, Winnipeg, Canada*

Abstract

The numerical search for the optimum shape of an aerofoil is of great interest for aircraft and turbomachine designers. Unfortunately, this process is very computationally intense and can require a large number of individual flow field simulations resulting in very long CPU run times. One of the core issues that the designer must deal with is how to describe the shape of the airfoil. Clearly, we can not treat the profile on a point by point basis as the problem would have an infinite number of degrees of freedom. Hence the typical practice is to resort to using a series of curves, such as polynomials and Bezier curves, to describe the profile. This typically reduces the number of degrees of freedom to a much smaller, manageable number. The influence of the selection of the parameterization on the optimization has received relatively little consideration to date. We can anticipate that some parameterizations will be less suitable for describing the profile shape and may result in slower convergence times.

Our paper will discuss a new airfoil parameterization, Bezier-PARSEC, that was developed to extend and improve the typical Bezier parameterization found in use. This parameterization was found to fit the known shape of a wide range of existing airfoil profiles as well as resulting in accelerated convergence for aerodynamic optimization using Differential Evolution. Our presentation will present the development and details of the Bezier-PARSEC parameterization and provide evidence that the parameterization is suitable and accelerates convergence.

Keywords: aerodynamic optimization, airfoil parameterization, Bezier curves.

WIT Transactions on The Built Environment, Vol 106, © 2009 WIT Press
www.witpress.com, ISSN 1743-3509 (on-line)
doi:10.2495/OP090181

1 Introduction

The profiles of most airfoils given in the literature are described by a tabulated vector of 50 to 80 coordinate pairs located on the profile surface as demonstrated by NACA compilation given by Abbot and von Doenhoff [1]. The practical demands of aerodynamic optimization on CPU time require us to search for an effective parameterization that contains fewer degrees of freedom. Four main objectives must be considered in developing an airfoil parameterization: minimizing the number of degrees of freedom as possible, the parameterization should be able to represent a wide range of existing airfoils, the parameters should be simple to formulate and impose, and finally, the parameterization should result in effective and efficient optimization.

The following work describes a new method of representing an airfoil that is an extension of a Bezier parameterization previously used by the authors [2] but uses aerodynamic parameters. These are similar to the PARSEC parameters developed by Sobieczky [3].

2 Bezier-PARSEC parameterization

A Bezier parameterization is determined by its control points which are physical points in the plane. The first and last control points are the initial and terminal point on the curve itself. However the other control points need not be on the curve even though they determine the shape of the curve. As such a Bezier parameterization of an airfoil has control points that are only indirectly determined by the underlying aerodynamics. It is desirable to have the parameters such as leading edge radius, trailing wedge angle, and so on that do have physical relevance to the flow, such as that demonstrated by Sobieczky [3, 4]. Oyama et al. [5] showed that this type of parameterization improved the robustness and convergence speed for aerodynamic optimization, which makes it more suitable for optimization using genetic algorithms. The superior performance of PARSEC parameterization is likely due to its ability to minimize epistasis – the nonlinear manner in which the objective function is dependent on the design parameters. Typically a reduction in nonlinear interaction of the parameters will enable the optimizer to converge more quickly.

This motivated the development of a combined Bezier-PARSEC (BP) parameterization to utilize the advantages of both the Bezier and PARSEC parameterizations. The BP parameters are aerodynamically oriented, and allow us to obtain the Bezier control points of four separate curves. These curves describe the leading and trailing portions of the camber line, and the leading and trailing portion of the thickness distributions. Each curve is ultimately a Bezier curve and will be denoted by BP ijkl, where i and j are the order of the leading and trailing thickness curve, and k and l are the order of the leading and trailing camber curves.

The parameters are: leading edge radius - r_{le}, trailing camber line angle - α_{te}, trailing wedge angle - β_{te}, trailing edge vertical displacement - z_{te}, leading edge direction γ_{le}, location of the camber crest - (x_c, y_c), curvature of the camber crest - κ_c, position of the thickness crest - (x_t, y_t), curvature of the thickness crest - κ_t, the half thickness of the trailing edge - dz_{te}, and several Bezier variable, b_0, b_2, b_8, b_{15} and b_{17}. The detailed development of the Bezier-PARSEC curves is given by Rogalsky [6] and only the results will be given here.

2.1 BP 3333 parameterization

A BP 3333 parameterization uses third degree Bezier curves for all four curves used to define the airfoil. A third degree Bezier curve is given parametrically by

$$x(u) = x_0(1-u)^3 + 3x_1u(1-u)^2 + 3x_2u^2(1-u) + x_3u^3,$$

and

$$y(u) = y_0(1-u)^3 + 3y_1u(1-u)^2 + 3y_2u^2(1-u) + y_3u^3,$$

where u is the parameter that runs from 0 at the beginning to 1 at its terminus. The BP 333 parameterization relies exclusively on the aerodynamic parameters – there are no free Bezier points in BP 3333. The x_i and y_i are the Bezier control points, which are computed from the following:

2.1.1 Leading edge thickness curve

The control points are given by

$x_0 = 0$ $\quad$ $y_o = 0$

$x_1 = 0$ $\quad$ $y_1 = 3\kappa_t(x_t - b_9)^2/2 + y_t$

$x_2 = b_9$ $\quad$ $y_2 = y_t$

$x_3 = x_t$ $\quad$ $y_3 = y_t$

The parameter b_9 the root of

$$27\kappa_t^2 b_9^4/4 - 27\kappa_t^2 x_t b_9^3 + \left(9\kappa_t y_t + 81\kappa_t^2 x_t^2/2\right)b_9^2$$
$$+\left(2r_{le} - 18\kappa_t x_t y_t - 27\kappa_t^2 x_t^3\right)b_9 + \left(3y_t^2 + 9\kappa_t x_t^2 y_t + 27\kappa_t^2 x_t^4/4\right) = 0$$

within the bounds given by

$$\max\left(0, x_t - \sqrt{-2y_t/3\kappa_t}\right) < b_9 < x_t.$$

2.1.2 Trailing edge thickness curve

The control points are given by

$x_0 = x_t$ $\quad$ $y_0 = y_t$

$x_1 = 2x_t - b_9$ $\quad$ $y_1 = y_t$

$x_2 = 1 + \left[dz_{te} - \left(3\kappa_t(x_t - b_9)^2/2 + y_t\right)\right]\cot(\beta_{te})$ $\quad$ $y_2 = 3\kappa_t(x_t - b_9)^2/2 + y_t$

$x_3 = 1$ $\quad$ $y_3 = dz_{te}$

2.1.3 Leading edge camber curve

The control points are given by

$x_0 = 0$ $\qquad$ $y_0 - 0$

$x_1 = b_1 \cot(\gamma_{le})$ $\qquad$ $y_1 = b_1$

$x_2 = x_c - \sqrt{2(b_1 - y_c)/3\kappa_c}$ $\qquad$ $y_2 = y_c$

$x_3 = x_c$ $\qquad$ $y_3 = y_c$

The parameter b_1 is computed from

$$b_1 = \left[16 + 3\kappa_c\left(\cot\gamma_{le} + \cot\alpha_{te}\right)\left(1 + z_{te}\cot\alpha_{te}\right)\right] / \left[3\kappa_c\left(\cot\gamma_{le} + \cot\alpha_{te}\right)\right]$$
$$\pm 4\sqrt{16 + 6\kappa_c\left(\cot\gamma_{le} + \cot\alpha_{te}\right)\left(1 - y_c\left(\cot\gamma_{le} + \cot\alpha_{te}\right) + z_{te}\cot\alpha_{te}\right)}$$

and must be within the following bounds

$$0 < b_1 < y_c.$$

2.1.4 Trailing edge camber curve

The control points are given by

$x_0 = x_c$ $\qquad$ $y_0 = y_c$

$x_1 = x_c + \sqrt{2(b_1 - y_c)/3\kappa_c}$ $\qquad$ $y_1 = y_c$

$x_2 = 1 + (z_{te} - b_1)\cot(\alpha_{te})$ $\qquad$ $y_2 = b_1$

$x_3 = 1$ $\qquad$ $y_3 = z_{te}$

2.2 BP 3434 Parameterization

A BP 3434 parameterization uses third degree Bezier curves for all of the leading edge curves and fourth degree Bezier curves for the both trailing curves used to define the airfoil. A fourth degree Bezier curve is given parametrically by

$$x(u) = x_0(1-u)^4 + 4x_1u(1-u)^3 + 6x_2u^2(1-u)^2 + 4x_3u^3(1-u) + x_4u^4,$$

and

$$y(u) = y_0(1-u)^4 + 4y_1u(1-u)^3 + 6y_2u^2(1-u)^2 + 4y_3u^3(1-u) + y_4u^4.$$

2.2.1 Leading edge thickness curve

The control points are given by

$x_0 = 0$ $\qquad$ $y_o = 0$

$x_1 = 0$ $\qquad$ $y_1 = b_8$

$x_2 = -3b_8^2/2r_{le}$ $\qquad$ $y_2 = y_t$

$x_3 = x_t$ $\qquad$ $y_3 = y_t$

Here the parameter b_1 is subject to the following restriction

$$0 < b_8 < \min\left(y_t, \sqrt{-2r_{le}x_t/3}\right).$$

2.2.2 Trailing edge thickness curve

The control points are given by

$x_0 = x_t \qquad y_0 = y_t$

$x_1 = \left(7x_t + 9b_8^2 / 2r_{le}\right)/4 \qquad y_1 = y_t$

$x_2 = 3x_t + 15b_8^2 / 4r_{le} \qquad y_2 = \left(y_t + b_8\right)/2$

$x_3 = b_{15} \qquad y_3 = dz_{te} + \left(1 - b_{15}\right)\tan\left(\beta_{te}\right)$

$x_4 = 1 \qquad y_4 = dz_{te}$

2.2.3 Leading edge camber curve

The control points are given by

$x_0 = 0 \qquad y_0 = 0$

$x_1 = b_0 \qquad y_1 = b_0 \tan\left(\gamma_{le}\right)$

$x_2 = b_2 \qquad y_2 = y_c$

$x_3 = x_c \qquad y_3 = y_c$

2.2.4 Trailing edge camber curve

The control points are given by

$x_0 = x_c \qquad y_0 = y_c$

$x_1 = \left(3x_c - y_c \cot\left(\gamma_{le}\right)\right)/2 \qquad y_1 = y_c$

$x_2 = \left(-8y_c \cot\left(\gamma_{le}\right) + 13x_c\right)/6 \qquad y_2 = 5y_c / 6$

$x_3 = b_{17} \qquad y_3 = z_{te} - \left(1 - b_{17}\right)\tan\left(\alpha_{te}\right)$

$x_4 = 1 \qquad y_4 = z_{te}$

The parameterizations assume that the length of the airfoil has been normalized to unit length.

2.3 Airfoil Representation

The parameterization method must be able tow represent a wide range of airfoils if it is to contribute to a robust design algorithm. The authors compared the ability of Bezier and BP methods to reproduce 63 known airfoils: 40 NACA symmetric and asymmetric airfoils [7], 15 Eppler airfoils [8], and 8 low-speed airfoils [9]. Data for each of the selected airfoils consisted of an ordered array of coordinates for each profile.

The curves were fit to the coordinates using a modified airfoil design code that used Differential Evolution to find the parameters that minimized the ℓ_2- error norm between the representation and the given data points. The DE algorithm was rand-to-best/1/exp with F=0.85, CR=1, and a population of 150 members. The maximum number of generations permitted was 500, however, in a few cases an additional run was used to obtain convergence. The convergence requirement was that the cost limit was less than 0.01. For a unit chord length airfoil this corresponds to an average deviation of approximately 8×10^{-4}.

The following discussion will be limited to a few representative airfoils due to limited space.

2.3.1 NACA Symmetric Airfoils

The Bezier, BP 3333 and BP 3434 approximations converged for all of the symmetric airfoils. Typically, fewer that 3,000 approximations, or 20 generations were required to obtain convergence, with one exception, the BP 3333 parameterization of the NACA 67_1-015. This indicates that matching the geometry of a known airfoil than for an inverse design using a Bezier parameterization [2].

The parameterization of the NACA 0008-34 airfoil will be presented in more detail. The curve fitting required the following; Bezier fit - 6,612 function evaluations with an average deviation of 1.06×10^{-4}, BP 3333 fit – 4,243 function evaluations with an average deviation of 1.24×10^{-4}, and BP 3434 fit – 6,180 function evaluations with an average deviation of 1.26×10^{-4}. Clearly, all of the parameterization methods could successfully approximate the shape of the symmetric profiles. The fit to the NACA 00008-34 is shown below, in Figure 1.

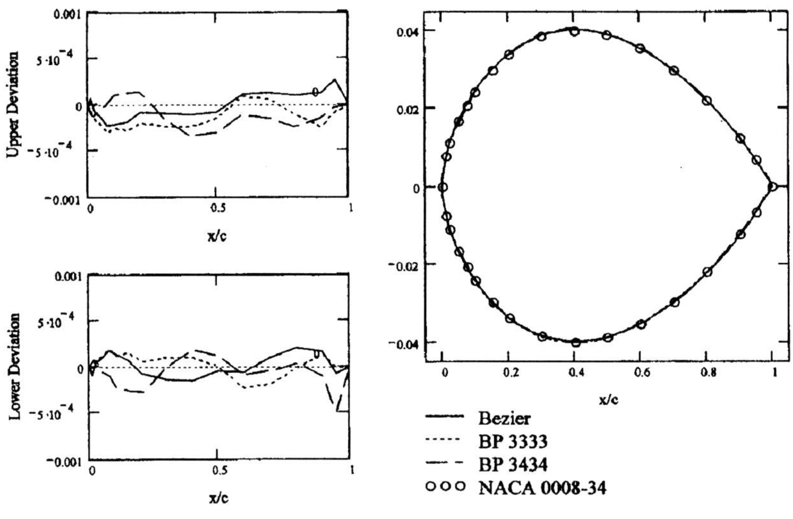

Figure 1: Representation of the NACA 0008-34 airfoil.

2.3.2 NACA asymmetric airfoils

The Bezier and BP 3434 were able to successfully reproduce each of the 20 asymmetric airfoils, with the BP 3333 failing to reproduce one, the NACA 744A315 to the required tolerance. The average number of function evaluations for the successful representations are as follows: Bezier fit – 3,302, BP 3333 - 3,071, and BP 3434 – 4,424. A successful fit for each method is shown below, in Figure 2, for the NACA 63_1-212 airfoil.

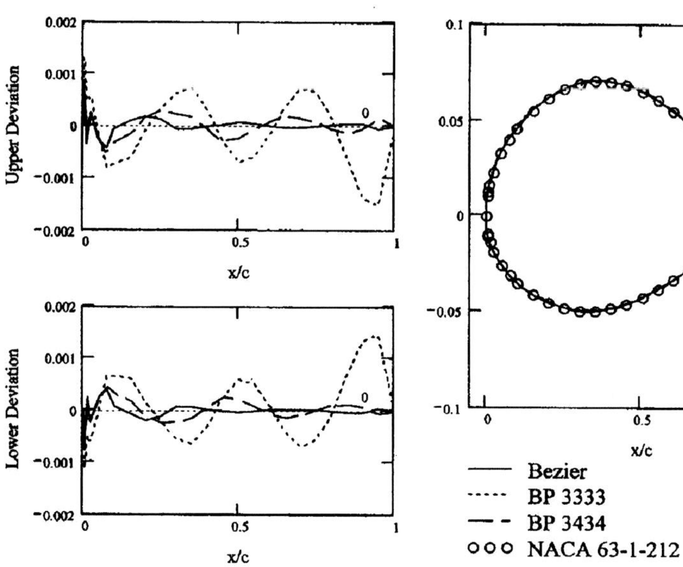

Figure 2: NACA 63-1-212 airfoil.

The failure of the BP 3333 parameterization for the NACA 747A315 was occurred near the trailing edge. This was primarily attributed to a lack of sufficient freedom on the trailing edge camber with minor issues with the trailing edge thickness profile. The NACA 747A315 has a sharp cusp at the trailing edge that is difficult to approximate. This was not an issue with the BP 3434 approximation as it has additional control points at the trailing edge.

2.3.3 Eppler airfoils

Eppler pursued the development of accurate theoretical methods to obtain inverse designs of airfoils with prescribed boundary layer characteristics. This has resulted in a catalogue of well established airfoils [8]. These airfoils represent a greater challenge due to their shape. The attempts to approximate these airfoils were less successful with the following success rates: Bezier – 73%, BP 3333 – 80%, and BP 3434 – 90%. The number of function evaluations for the successful approximations were: Bezier – 3,316, BP 3333 – 3,570, and BP 3434 – 4,870. The resulting approximations for the E 266 airfoil are shown below in Figure 3.

The reasons for the failure to approximate the Eppler airfoils to the required accuracy are numerous, and include a lack of degrees of freedom, difficulty with airfoils that have a positive leading edge and negative trailing edge directions, and finite trailing edge thickness to name a few issues.

2.3.4 Low-speed airfoils

All of the methods were able to reproduce the selected set of low-speed airfoils to the desired tolerance. The average number of function evaluations required to

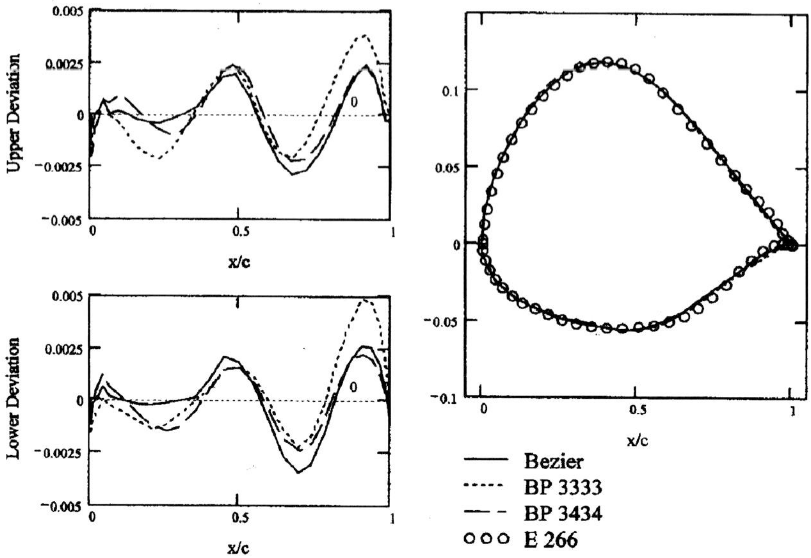

Figure 3: Eppler E 266 airfoil.

fit the airfoils were: Bezier – 5,483, BP 3333 – 8,834, and BP 3434 – 9,994. A representative fit for the low-speed airfoils is given for the FX 74-CL5-140 MOD airfoil shown in Figure 4.

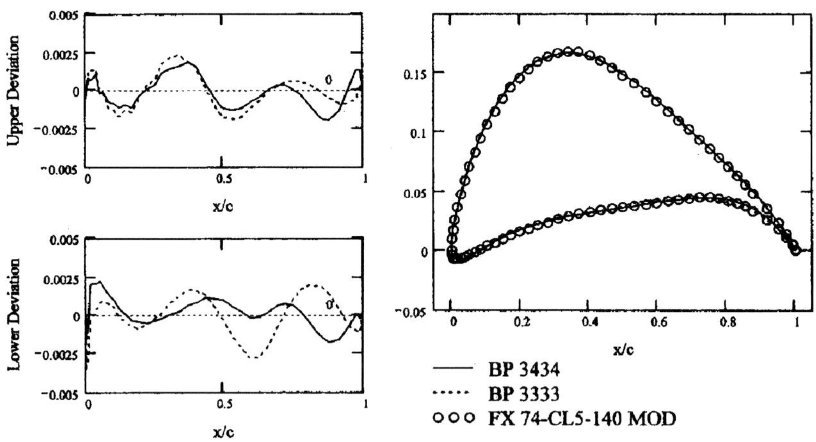

Figure 4: Low-speed FX 74-CL5-140 MOD airfoil.

2.4 Summary

Each of the three parameterization methods was able to represent a high percentage of the sample airfoils. Of the 63 airfoils, the Bezier parameterization reproduced 58, BP 3333 reproduced 59, and BP 3434 reproduced 62. In most cases the approximation required less than 10,000 function evaluations. Typically, the Bezier approximations required the fewest function evaluations and the BP 3434 approximations required the most.

The biggest limitation of the Bezier approximations is the discontinuous second derivative at the camber and thickness crests which can result in inaccurate flow field simulations. A less significant difficult is the zero trailing edge thickness which is a problem for some airfoils such as the Eppler E 863.

The BP 3333 approximation has the least control point freedom which can be beneficial in some cases such as for the Eppler E 417. Two classes of airfoil cannot be successfully represented by BP 3333; those with a radical change in the trailing edge camber curve and those with a camber curve that dips below the x-axis.

The BP 3434 approximation is the most robust due to the additional control points at the trailing edge. This can be a problem in that it can result in the camber crest being pushed to far back or premature convergence with the shapes with sharp edges of incorrect trailing edge directions.

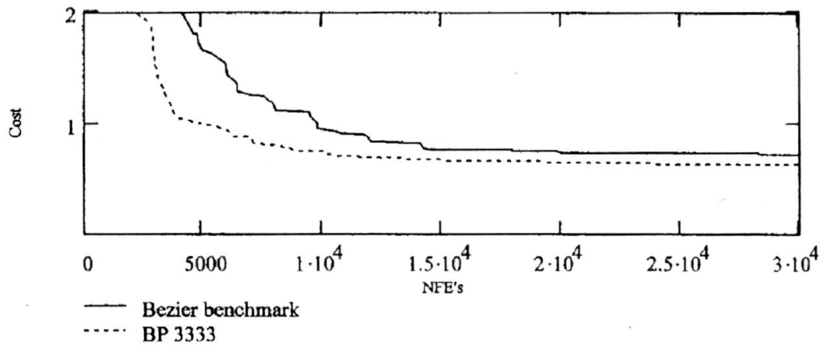

Figure 5: Effect on convergence.

3 Effect of parameterization on design speed

The previous discussion showed that all three parameterization methods can represent a broad range of airfoils. The next issue of concern is the affect of the parameterization on the robustness and rate of convergence for an inverse design problem. The focus is narrowed to a comparison between the Bezier and BP 3333 parameterizations. The results of an optimization to a 112° cambered turbine blade will be presented here. The optimization was based on finding the target airfoil that best matches the known, theoretical, pressure distribution on an airfoil. This allows us to determine the robustness and convergence properties of the optimization process, while insuring that the optimum has been reached.

3.1 Convergence for a 112° cambered turbine blade

The convergence rate for the Bezier and BP 3333 parameterizations are shown below. Clearly the BP 3333 parameterization results in a more rapid convergence to the solution than the Bezier parameterization for this case, and was also found to be the case for a Liebeck design. However, similar convergence histories for Bezier and BP 3333 parameterizations occurred for a C4/70/C50 turbine blade target. Additionally, the BP 3333 approximation did a better job of converging to the target airfoil than the Bezier approximation.

4 Conclusions

The three parameterization methods are all capable of representing a wide range of airfoil shapes with a slight edge going to BP 3434. However, the BP 3333 parameterization performed better, and certainly no worse than the Bezier parameterization for inverse design.

References

[1] Abbot, I.H., and von Doenhoff, A.E., *Theory of Wing Sections*, Dover Publications, New York, 1959.
[2] Rogalsky, T., Derksen, R.W., and Kocabiyik, S., Differential Evolution in Aerodynamic Optimization, *Canadian Aeronautics and Space Journal*, **46(4)**, pp. 183-190, 2000.
[3] Sobieczky, H., Parametric Airfoils and Wings. *Recent Developments of Aerodynamic Design Methodologies*, ed. Fujii, K. and Dulikravich, G.S., Friedr. Vieweg & Sohn Verlagsgesellschaft mbH, Braunschweig/Wiesbaden, pp. 71-87, 1999.
[4] Sobiczky, H., Manual Aerodynamic Optimization of an Oblique Flying Wing, 36th Aerospace Sciences Meeting & Exhibit, Reno, NV, *AIAA Paper* 98-0598, 1998.
[5] Oyama, A., Obayashi, S., and Nakahashi, K. Fractional Factorial Design of genetic coding for aerodynamic optimization, *AIAA Paper* 99-3298, 1999.
[6] Rogalsky, T. *Acceleration of Differential Evolution for Aerodynamic Design*, Ph.D. Thesis, University of Manitoba, 2004.
[7] Abbot, I.H., and von Doenhoff, A.E., *Theory of Wing Sections*, Dover, New York, 1959.
[8] Eppler, R., *Airfoil Design and Data*, Springer-Verlag, Berlin, 1990
[9] Selig, M.S., Guglielmo, J.J., Broeren, A.P. and Giguère. P., *Summary of Low-Speed Airfoil Data*, Vol. 1, SoarTech Publications, Virginia Beach, 1995.

Development of a novel Electro Active Polymer (EAP) actuator for driving the wings of flapping micro air vehicle

S. C. Burgess, C. S. Ling, A. Conn, S. Araromi, J. Wang & R. Vaidyanathan
Department of Mechanical Engineering, University of Bristol, UK

Abstract

This paper describes the development and testing of an EAP (electroactive polymer) actuator for driving the wings of an MAV (micro air vehicle). Creating the drive system for a flapping MAV is extremely challenging because of the required combined bending/twisting of the wing during flapping. Silicone-based DEs (dielectric elastomer) were used for the actuator material. In addition, the EAP actuators were operated at resonant frequency which is an important means of producing optimal flight performance with reduced power consumption. The paper presents the optimisation of the structural and operational performance of the EAP actuator. The paper also presents the test results of the actuator driving an actual MAV wing.
Keywords: electroactive polymer, micro air vehicle, resonant frequency.

1 Introduction

MAVs that use insect-like flapping for lift have the potential for achieving advanced flight capabilities such as silent operation; high payload; controlled hovering; and high manoeuvrability. Such flight performance is required for reconnaissance in confined places like buildings, urban areas, tunnels and caves. Insect-like flapping involves simultaneously flapping and twisting of the wings and the production of unsteady aerodynamic effects such as the leading edge vortex [1]. These unsteady aerodynamic effects are responsible for the high lift generated in insect flight. Creating a controllable MAV is extremely challenging because of the required combined bending/twisting wing movement. EAP

WIT Transactions on The Built Environment, Vol 106, © 2009 WIT Press
www.witpress.com, ISSN 1743-3509 (on-line)
doi:10.2495/OP090191

actuators are well suited for MAVs because they are solid state actuators which have high strain outputs. They are an emerging technology and have the advantage that they are mechanically simple, compact and light-weight for many applications. In particular, they often do not require gearing in a drive system because of the adaptability of configuration and high strain output. Many existing flapping wing MAV designs are limited in performance because they use conventional constrained mechanisms like crank-rockers [2–5].

A key advantage of having a linear drive to produce the wing reciprocating motion is the ability to oscillate at resonant frequency. This can save energy by taking away the need to accelerate and decelerate the wings. The work by Goldfarb and colleagues at Vanderbilt University demonstrated the use of resonant flapping through the excitation of the torsional and bending natural frequencies using piezoelectric ceramic benders [6]. Similar resonant flapping demonstrators have also been reported in [7–9]. Another important advantage of linear actuators is that it is easier to adjust control surfaces such as the angle of attack of an individual wing as they are suited to driving under-constrained mechanisms.

2 Optimisation of flapping wing drive system

A schematic of a typical rotary drive actuation system such as a crank-rocker is shown in Fig. 1. Compliances and backlash in the drive system lead to shock loads and induced time lag from activation to output. Energy is also wasted on accelerating the masses of the various transmission components, effectively reducing efficiency.

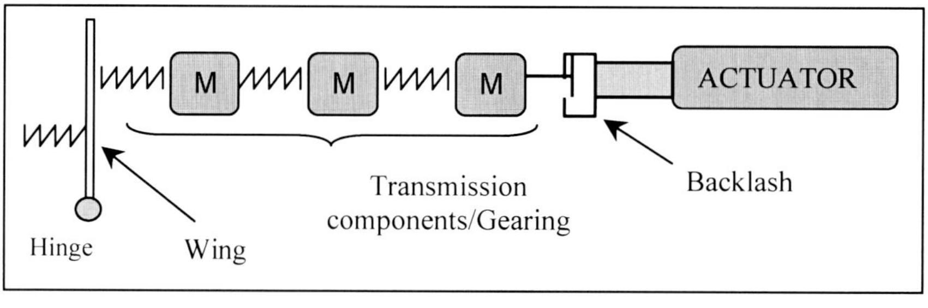

Figure 1: Typical rotary drive actuation system.

A schematic of the EAP linear actuation system developed at Bristol University is shown in Fig.2. The system is inherently more stiff and controllable.

An EAP or DE actuator is inherently well-suited for resonant operation because it is solid-state with no mechanisms and linkages. The DE bender designed in this project (Fig. 2) behaves like a cantilever beam (representing the shim substrate) with a uniform load (representing the elastomer film stack) acting across its length. The natural frequency for a cantilever beam uniform load can be calculated using equation (1):

$$f_n = \frac{K_n}{2\pi}\sqrt{\frac{EIg}{wl^4}} \tag{1}$$

where w is the force per unit length of the uniform load (including shim substrate weight), l is the bender free length, E is the Young's modulus of the beam, I is the moment of inertia, g is the acceleration due to gravity and K_n is 3.52 for the first mode of resonance.

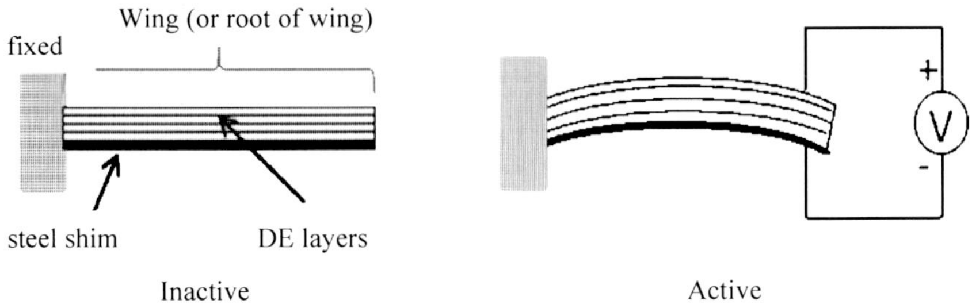

Figure 2: EAP bender actuator developed at Bristol University.

For the DE bender configuration, the elastic modulus of the steel shim substrate is assumed to be E = 200GPa and the inertia I = (shim thickness)^3 × (shim width)/12, where the shim thickness = 50μm and shim width = 10mm. The bender free length, l, is 60mm and the uniform load, w, is approximated as the actuator weight (equal to 39.2 N) divided by l. Therefore, from the equation above, the natural frequency, f_n, is approximately equal to 2.8 Hz.

This closely matches the observed value during testing, which was approximately equal to 2.7 Hz. Note that from equation (1) the actuator resonant frequency will increase with increase substrate moment of area and Young's modulus, however this would also decrease strain output. Therefore, there is a compromise between resonant frequency and strain output.

3 Optimisation of actuator design for flapping

3.1 Material properties of DEs

Most existing linear actuators are not suitable for MAVs because they require either additional strain amplification devices (e.g. piezo ceramic, conductive/electrorestrictive polymers) or have a limited power output (e.g. shape memory alloy, solenoid). However, DE actuators produce high strain at high power output. When an electric field is passed through the thin polymer film, sandwiched between two compliant electrodes, physical deformation takes place due to the effect of opposite charges attracting and like charges repelling. This causes the DE layer to be 'squeezed' in the transverse direction and expand in the planar direction.

Silicone based Dow Corning HSIII and NuSil CF19-2186 have maximum volume specific energy densities of 160kJ/m^3 and 220kJ/m^3 respectively, and acrylic type elastomer 3M VHB 4910 has a maximum volume-specific energy of 3400kJ/m^3 (estimated) [11]. For comparison, muscle has a maximum volume-specify energy of 70kJ/m^3 [12]. Table 1 presents a comparison of properties of muscle and DEs. The table shows that EAPs have high stress and strain outputs.

Table 1: Comparison of performance of muscle and DEs [13].

Property	**Mammalian muscle**	**Dielectric elastomers**
Strain (%)	20	10 to >100
Stress (MPa)	0.1-0.35	0.1-2 typical, max 3-9
Work density (kJ/m^3)	8-40	10-150 typical, max 3400
Density (kg/m^3)	1037	~ 1000
Strain rate (%/s)	>50	450 in VHB, 34000 in silicone
Continuous power (W)	50- 280	~ 500
Electromech. coupling (%)	< 40 efficiency	30 typical, max of 90
Cycle life	>10^9	10^6 @ 50% strain
Modulus (MPa)	10-60	0.1-3
Voltage (V)	< 1V	>1kV

3.2 Governing equations for DEs

Using a simple electrostatic model, the pressure, p, exerted by the electrodes on the elastomer film is given in eqn. (2).

$$p = e_r e_0 T^2 = e_r e_0 (V/h)^2 \tag{2}$$

where T is the applied electric field, V is the applied voltage, h is the elastomer film thickness, e_r is the relative dielectric constant of the elastomer and e_0 is the permittivity of free space. From eqn. (2), the effective pressure, p, is essentially related to the force output of the DE actuator. For a given effective pressure, the higher the dielectric constant (e_r) and the thinner the elastomer film is, the lower the electric field (V) is needed to achieve this force. Assuming the elastomer film is unloaded and unconstrained, the thickness strain, ε_z, that results from p is given by:

$$\varepsilon_z = \frac{-p}{E} = \frac{-e_r e_0 (V/h)^2}{E} \tag{3}$$

where E is the Young's Modulus of the elastomer.

3.3 The bender DE actuator developed at Bristol University

The bender produced in this work consists of a thin passive substrate (steel shim of 50μm in thickness) with layers of active DE with carbon electrode on the top surface. The electrode layers were stacked with alternating polarity, creating the active layers under activation. Matching polarity active layers were connected electrically using pre-cut aluminium strips, while self adhesive copper tape was used in the final stage to provide the electrical interface for the high voltage power supply. The composite beam bends when the EAP material is activated.

An in-house fabrication technique has been developed for production of thin silicone films. Two types of silicone DEs (BJB TC-5005, Dow Corning Silastic 3481) and one type acrylic DE (3M VHB 4910) have been acquired. The treatment and fabrication technique for each of these DEs are different. Films with thicknesses of 100, 200 and 500μm have been produced by rolling the K-bar on the BJB TC-5005 silicone mixture. A similar technique, however, was found to be inappropriate for the Dow Corning Silastic 3481 (HSIII equivalent) which has higher surface tension. The mixture of the Silastic 3481 silicone (part A) and the curing agent (part B) has to be 'dissolved' in the solvent, and sprayed onto the steel substrate using an air brush. The 3M VHB 4910 acrylic elastomer, on the other hand, comes pre-fabricated with minimum thickness of 500μm and thus requires pre-straining biaxially before the compliant electrode can be applied.

3.4 DE bender design considerations

The DE actuator requires a high field strength, typically in the range of 50-200MV/m. The recent miniaturisation of step-up voltage converters (e.g. EMCO Q50-5) make them convenient and feasible for use in MAVs. The requirement for minimising the film thickness to reduce voltage requirements, coupled with the elastomer's finite breakdown voltage, means that a stacked configuration is necessary to achieve sufficient power output.

It is also essential to have highly compliant electrodes which maintain good conductivity under high strain rates. Several possible conductive materials have been identified for the electrode of the DE bender. These are grease doped with carbon or silver, silicone and carbon powder mix, Metal rubber™ (Nanosonic Inc., USA) and spluttered gold layers. The use of Metal rubber™ is limited by its high cost. Gold is difficult to apply because gaps have to be introduced during spluttering due to its high stiffness. Conductive grease is not suitable for stacking elastomers, as grease does not provide adhesion between the layers and that causes the layers to delaminate. A mixture of the silicone BJB TC-5005 with carbon was found to be the most appropriate candidate for the stack elastomer configuration.

The bender configuration is mass efficient (without hinges) and the passive substrate (steel shim) provides elastic storage. The multi-layered bender configuration pursued in this work benefits from the high power output of stacked films, combined with the high strain output and an elastic storage capability, make the choice of such configuration an optimal design.

4 DE bender and flapping wing configurations

4.1 Single degree of freedom flapping

Configuration 1
The first and most basic configuration attempted to provide flapping using a single degree of freedom (DOF) multi-layered DE actuator with a scaled model wing attached rigidly to the actuator tip. The actuator produced ample force to flap the wing and overcome inertial forces. The actuation system was also able to be driven at resonant frequency and a maximum sweep angle of 40° – 50° was observed in testing using an 8-layer actuator stack.
Configuration 2
To achieve strain amplification a simple lever mechanism was constructed. A carbon fibre rod was attached perpendicular to the protruding end of the leading edge wing spar. The rod was inserted into an aluminium tube with an inner diameter larger than that of the rod, therefore allowing it to rotate freely, thus facilitating the pivot for the lever mechanism as shown in Fig. 3.

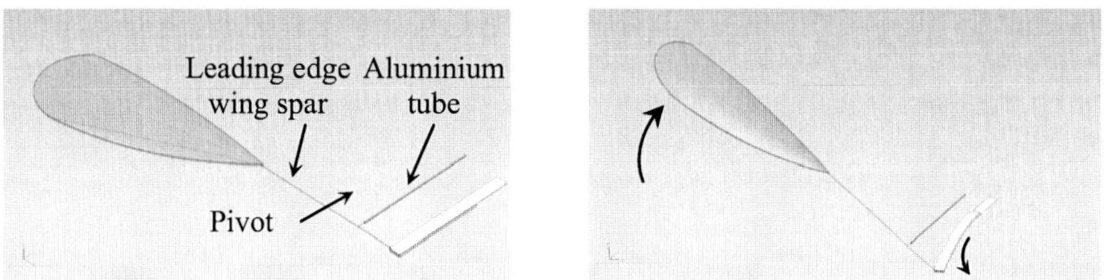

Figure 3: EAP actuator with lever mechanism for strain amplification.

The prototype built was capable of being driven at resonance. Moreover, the moment produced by the mass of the wings provided some pre-tensioning on the DE actuator and the actuator motion worked with the natural rotation produced by the induced moment arm of the wing's mass. This resistance is primarily due to friction forces at the interface which may be overcome by rigidly attaching the wing spar to the actuator tip. However, the motion of the spar relative to the actuator is planar due to the curved profile of the activated actuator; hence the design of such a linkage is nontrivial.

4.2 Flapping with passive twist

The two configurations investigated above were adapted to allow for passive twisting of the wing about the length axis of the leading edge wing spar. The twisting is a result of aerodynamic forces and wing inertia. In insects, such as the *Libellula pulchella* dragonfly, these forces are sufficient to provide the wing pitching necessary for steady flight, without the aid of muscles [14].
Configuration 3
A design was constructed which allowed for twisting by placing the protruding end of the leading edge wing spar into the small length of aluminium tubing

rigidly attached to the tip of the actuator. The mechanism is similar to that used to provide the pivot in Configuration 2, allowing for free rotation about the leading edge of the spar axis. A pitching angle of approximately ±5° was observed with this design in simple flapping test.

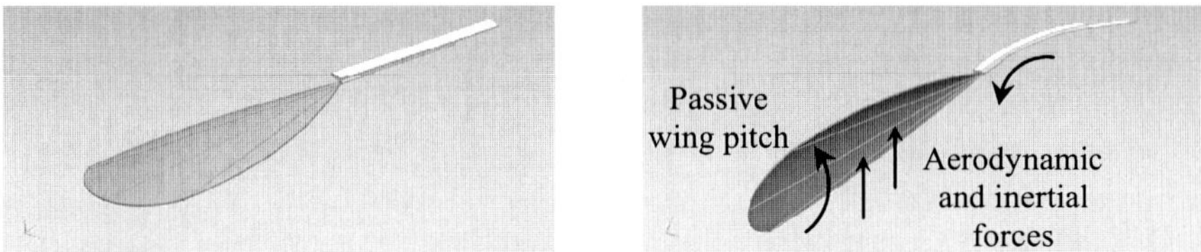

Figure 4: Wing flapping with passive twist.

Configuration 4
Configuration 3 can be adapted to incorporate the strain amplification mechanism utilised in Configuration 2, thus generating a greater sweep angle whilst maintaining passive pitching capabilities.

4.3 Wing flapping with controlled twist

Although there is evidence to show that passive wing pitching induced by aerodynamic and inertial forces may be sufficient for steady state flight applications, the ideal case would be to have dynamic control over wing pitching angle in addition to simple flapping motion.
Configuration 5
Configuration 5 was developed using a modified version of Configuration 3 with the addition of a second EAP actuator in parallel with the first. Initially, the two actuators were arranged so that they would act in opposite direction when activated, causing the wing to twist as a result.

Figure 5: Flapping with controlled twist.

While this design is rather promising, the resistance to motion arising from frictional forces resulted in very little pitching being produced. The actuators also appeared to work against each other and produced no flapping. In an attempt to rectify this problem, the actuators were repositioned so that they acted in the same direction when activated, with one of two actuators given a slightly reduced voltage input to create a difference in sweep angle. In this case, friction in the system caused the actuators to behave as though they were rigidly attached via the pitching rod and therefore no twisting was produced.

4.4 Configuration summary

Proof of concept flapping wing designs have been developed using DE benders. Configuration 3 shows good potential for further development with minimal design complexity. The limiting factor in the more complex designs was friction which in some cases, greatly reduced strain output. Moreover, future DE actuator designs will be investigated for two degree of freedom actuation. Such an actuator would also increase dynamic control of the wing motion whilst maintaining design simplicity.

5 Experimental results

To evaluate the performance of the DE bender, a simple strain test to measure the bender tip deflection was used. A number of stacked benders, comprising of 4, 8 and 12 layers of Dow Corning Silastic 3481 were fabricated. Each of these DE benders was placed inside a high-voltage isolation box with an EMCO F12-TR high voltage converter. The benders were clamped on one end using a mounting block, while graph paper with 2mm grid was placed directly under the bender. The supply voltage to the bender was incrementally increased from 0-7.2kV, and the tip deflection for each of these benders was captured using a digital camera mounted perpendicularly to the 2mm grid. In between the measurements, the power supply was switched off to allow the bender to return to its original inactive position. This process was repeated for each bender and Fig. 6 shows the inactive and active state of the 12-layer bender.

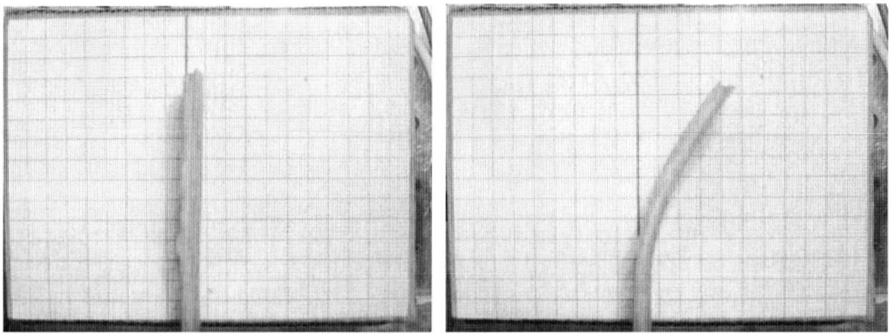

Figure 6: Dow Corning Silastic 3481 12-layer stacked bender when it is inactive (left). Tip displacement for the bender for a voltage of 6.4kv (right).

5.1 Tip displacement measurements

The tip deflection measurements for the 4 and 8 and 12-layer are shown in Fig. 7. The 4 and 8 and 12-layer bender produce maximum tip deflection of 2.4mm 7.3mm and 19.2mm respectively.

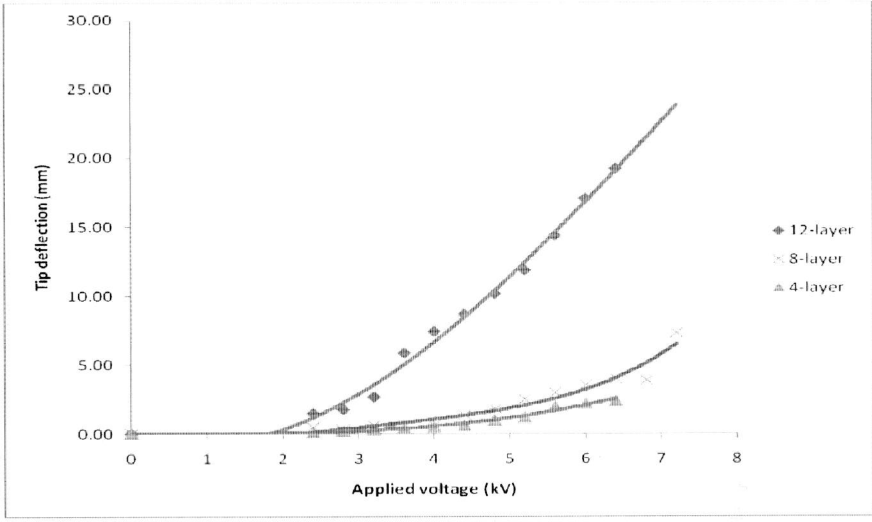

Figure 7: Tip displacement vs input voltage (kV) for 4, 8 and 12-layer DE bender.

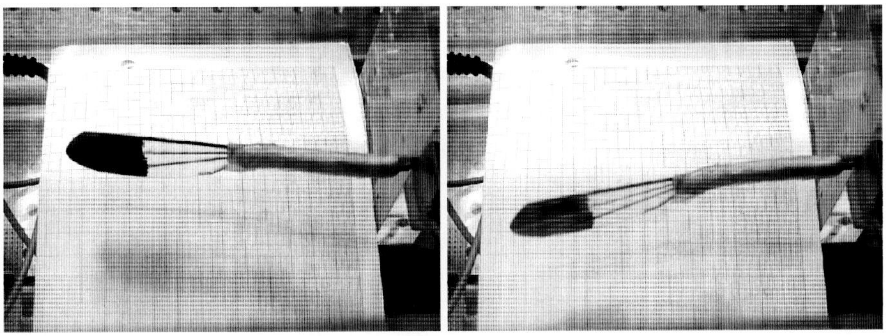

Figure 8: DE bender with wing attached during resonant operation.

5.2 DE bender with wing in resonant operation

A wing with spars made of thin carbon fibre rods and a Mylar membrane was attached to the end tip of the bender (same arrangement as Configuration 1 in Section 4.1). The actuator was driven by a square wave input with an adjustable frequency in an attempt to resonate the wing. A resonant frequency of approximately 3Hz was identified for a 12-layer DE stacked bender (Fig. 8). When the voltage is turned off, the energy stored in the substrate (due to its stiffness) the bender-wing arrangement returns naturally to its original position. The momentum acquired during the upstroke cycle causes the wing to travel further than its original position, resulting in a reciprocating wing motion. When the actuator stack is oscillating at its natural frequency, the resonant operation has the effect of increasing stroke amplitude whilst minimising inertial energetic expenditure. The moment of inertia or stiffness of the bender will have to be increased in order to achieve the desired 20-30Hz operation frequency for MAV.

6 Conclusion

The high volume-specific power density and high strain performance of DEs make them an inherently good choice of actuator for flapping wing MAVs. The ability to resonate EAPs as a solid state structure gives potential for high efficiency of operation. Resonance has been demonstrated at around 3 Hz for the bender actuator developed in this project.

Various configurations of bender actuators were designed, fabricated and tested. Configuration 3 was the simplest and allowed for flapping and passive wing pitching which is induced by aerodynamic and inertial forces. Preliminary test results show that the DE actuator exhibits potential for producing an optimally design flapping mechanism. Further improvements of the actuator configuration and fabrication techniques are needed to achieve actual flight of an MAV.

References

[1] Sane, S.P., The aerodynamics of insect flight. *Journal of Experimental Biology*, **206**, pp. 4191-4208, 2003.

[2] Conn, A. T., Burgess, S. C., Ling, C. S. & Vaidyanathan, R., The design optimisation of an insect- inspired micro air vehicle. *Journal of Design & Nature*, WIT, **3(1)**, pp 12 - 27, 2008.

[3] Conn, A. T., Burgess, S. C., and Ling, C. S., Design of a parallel crank-rocker flapping mechanism for insect-inspired micro air vehicles. *Proc. IMechE, Part C: Journal of Mechanical Engineering Science*, **221**, Part C, pp. 1211-1222, 2007.

[4] Jones, K.D. and Platzer, M.F., Bio-inspired design of flapping wing micro air vehicles – an engineer's perspective. *44th Aerospace Sciences Meeting and Exhibit*, 9-12 Jan, Reno, 2006.

[5] Lentink, D., Bradshaw N.L. and Jongerius, S.R, Novel micro aircraft inspired by insect flight. *Comparative Biochemistry and Physiology Part A*, **146A**, Supplement 1 (April 2007), pp. S133-S134, 2007.

[6] Cox, A., Monopoli, D., Cveticanin, D., Goldfarb, M., and Garcia, E., The development of elastodynamic components for piezoelectrically actuated flapping micro-air vehicles. *Journal of Intelligent Material Systems and Structures*, **13**, pp. 611-615, 2002.

[7] Luc-Bouhali, A. Progress of the REMANTA project on MAV with flapping wings. *European Micro Air Vehicle (EMAV) Conference and Flight Competition*, Jul 25-26, Braunschweig, 2006.

[8] Syaifuddin, M., Park, H.C. and Goo, N.S., Design and evaluation of a LIPCA-actuated flapping device. *Smart Materials and Structures*, **15**, pp. 1225-1230, 2006

[9] Wood, R.J., Steltz, E. and Fearing, R.S., Optimal energy density piezoelectric bending actuators. Sensors and Actuators A: Physical, **119**, pp. 476-488, 2005.
[10] Avadhanula, S., Wood, R.J., Steltz, E., Yan, J. and Fearing, R.S., Lift force improvements for the micromechanical flying insect. IEEE International Conference on Intelligent Robots and Systems, Oct 28-30, Las Vegas, 2003.
[11] Pelrine, R., Kornbluh, R., Joseph, J., Heydt, R., Pei, Q. and Chiba, S., High-field deformation of elastomeric dielectrics for actuators. Materials Science and Engineering C, **11**, pp. 89-100, 2000.
[12] Bar-Cohen, Y, Electroactive Polymer (EAP) Actuators as Artificial Muscle, 2nd edn. SPIE, Bellingham, WA, 2004.
[13] Pei, Q., Rosenthal, M. A., Pelrine, R., Stanford, S. and Kornbluh, R. D., Multifunctional electroelastomer roll actuators and their application for biomimetic walking robots [5051-31]. Proceeding of SPIE, 5051, pp. 281–290, 2003.
[14] Bergou A.J., Xu, S., Wang, Z. J. (2007). Passive wing pitch reversal in insect flight. *Journal of Fluid dynamics*, **59**, pp. 321-337, 2007.

Section 6
Optimisation in dam and soil engineering

A note on the search for 'optimal' shapes of arch dams

M. Fanelli[1] & A. Fanelli[2]
[1]*Private Consultant, Former Director of the Centre for Hydraulic & Structural Research of ENEL, Italy*
[2]*Private Consultant, Italy*

Abstract

The 'shape resistance' of an arch dam affords a concrete volume substantially smaller than that of a gravity dam of equal height, at the expense of more stringent requirements for concrete quality, formwork, foundation treatments, geometric tracing and erection constraints. Minimisation of the concrete volume or, more generally, of total costs is pursued by 'optimising' the shape under a set of constraints. This requires defining a 'structural model', a set of 'shape parameters', the 'constraints' under which optimisation is to be achieved, the 'design external load(s)', an 'objective function' and an 'algorithmic procedure'. Usually the number of shape parameters ranges from a dozen to 40 or more, to leave wide freedom of adaptation of the final shape to loads and constraints, and a unique distribution of design loads is assumed; hence heavy analytical-numerical procedures leading to a unique 'optimal' shape. As the dam is called to withstand a large variety of external actions, the shape thus obtained cannot be 'optimal' for all cases; therefore the rationale for complex optimisation models may be questioned. The Authors claim that the number of shape parameters can be kept very small (from three to a few units) by adopting a simplified model complying from the start with the main constraints. This claim is supported by past experience: the shapes of arch dams designed in this way were found, indeed, to be very close to those obtained by more 'rigorous' approaches.

Keywords: arch dams, shape optimisation, shape parameters, constraints, simplified models.

doi:10.2495/OP090201

1 Introduction

State of the art does not allow 'direct' design of an arch dam. Experience, trial-and-error procedures, algorithmic 'optimisation' methods are the tools currently available in order to define a structurally adequate shape avoiding unnecessary costs. In principle only an infinite number of geometrical parameters could accurately define a shape which is totally unknown to start with. Even if in actual practice a search for such an extreme accuracy is not warranted for, in most of the literature a large number of geometrical parameters (between a dozen and several tens) is currently deemed necessary to leave adequate 'freedom' to the search strategy. The 'objective function' to be minimized (under the given constraints) is usually defined as the overall cost of erection, of which the main components are proportional to **i)** the volume of concrete and **ii)** the surface and volume of the excavations and foundation works (sometimes only **i)** is considered). The constraints under which 'shape optimisation' is to be sought relate mainly to geometry (adaptation to local topography/geology), to the building schedule (e.g. the choice to erect the individual monoliths independently from each other, i.e. avoiding the necessity of coordinated pouring and progressive joint grouting.) and to structural adequacy, i.e. the capacity to withstand a given set of loads without 'failure' (for the definition of 'failure' see the ICOLD glossary). As to the latter, 'admissible stress' resistance criteria are assumed; but the constrained stresses are not explicit functions of the shape parameters, and to find the minimum of a function of a large number of variables under implicit constraints turns out to be a computationally expensive task (FE stress analysis is currently used). Moreover, the 'precision' of the solution pursued by such a search can be delusory, insofar as the purported 'optimal' shape is obtained in relation to a definite (unique) set of loads, whereas the 'real' dam is called to face a large variety of different load scenarios. Simplified optimisation methods preserving the essential features of the problem and allowing for judicious intervention of the designer's experience may therefore be proposed (if proven valid in practice) as more expeditious approaches. The present note suggests that a pared-down model based on suitably chosen variables might avoid FE computing and reduce the degrees of freedom of the search to only three/four.

2 Generalities

Arch dam structural analysis encompasses two main classes of problems: **i)** analysis of existing (or anyway pre-defined) structures and **ii)** design of dams still to be built. The 'brute force' approach to the second problem would proceed through a 'trial and error' procedure. To wit, a large number of 'compatible' geometric shapes (i.e. shapes compatible with the local topography/geology.) (generated either by a systematic scanning of the possibilities or by the designer's intuition and experience) would be analysed by the methods used for the first class of problems (**i**); a screening of the results would then identify the solution(s) compatible with the constraints and presenting as little costs as

possible. While there is nothing wrong in this way of proceeding, more satisfactory lines of approach are devised, i.e. ones that embody greater objectivity (as compared with the experienced engineer's intuitive approach) avoiding at the same time excessive computational costs (as compared with the brute-force trial-and-error approach). The choice among the many options for such methodologies should be guided by the criterion that a cunning selection of the main variables is the best premise to identifying an effective mathematical model of the problem in hand (J.H. POINCARE'). The approach herein presented starts from the classical RITTER idea of splitting the design load into two fractions: one carried by the bending stiffness of the vertical cantilevers and the remaining part carried by the axial stiffness of the horizontal arches. The stress distribution caused by dead weight is considered apart, under the assumption that the individual monoliths are so designed as to avoid the necessity of coordinated pouring schedule and intermediate grouting of joints. This 'self-standing' condition implies that the excentricity of the vertical dead-weight load acting on each horizontal section of an isolated monolith is everywhere contained within 1/6 of the width of the section. In order to take maximum advantage, under the design load, of the dead-weight pre-stressing thus generated, it is necessary that this excentricity be directed downstream in the upper part of the cantilever and upstream in the lower part. The excentricity diagram under the design load is qualitatively reversed (i.e. directed upstream of the section centre in the upper part of the cantilever and downstream in the lower part), and to avoid in-service tensile vertical stresses it should also remain contained in either zone within 1/6 of the local section width. These two constraints are inherently embodied in the basic model herein adopted, so that it is unnecessary to check their compliance 'a posteriori'. The 'pilot' element of the design-load splitting becomes now the crown cantilever (In the traditional RITTER approach the pilot elements are the arches, whose radius is pre-defined, and the load fraction not absorbed by the arches must be taken up by the cantilevers. Therefore there is no 'a priori' assurance that vertical tensile stresses are avoided, but one has to check this condition 'a posteriori' and to accept or modify the design on the basis of the check results.), which is assumed to take up just the load fraction it can carry without tensile stresses; the remaining fraction taken up by the arches is then determined by difference with the design load, and the crown radii of curvature are accordingly defined. In the simplified presentation of our linear model (next §**3**) the following assumptions are made:
-a symmetrical configuration is assumed, and only the equilibrium of the central cantilever is explicitly formulated,
-the design load is assumed to consist of horizontal actions (external pressures on the upstream face); uplift is disregarded,
-the mutual actions exchanged between arches and central cantilever are likewise assumed to be horizontal; torsional stiffness is neglected,
-the foundation compliance is represented through the VOGT coefficients,
-thermal loads are not included in the formal load-splitting (they can be taken account of in the final stress checks); dynamic actions are replaced by pseudo-static ones according to the WESTERGAARD theory.

Extensive prior experience of the Authors in design (or in back-analysis of in-service behaviour) of large arch dams gave empirical evidence that these approximations do not significantly distort the structural modelling. Besides, the uncertainties about the input data and the necessarily sketchy character of the design load are not logically consistent – always in the Authors' opinion – with the use of very sophisticated models.

3 Basic equations, parameter fields of variation, search strategy

The following presentation implies reference to a 'reduced scale model' of the dam in which the actual height H is brought down to 1 m and the specific weights of water and concrete are increased in the ratio $H/1$ in order to preserve the magnitude of pressures and stresses. Symbol ζ denotes the ratio z/H, where the vertical coordinate z is measured downwards from the dam crest, and H is the actual height of the dam. Three functions of ζ define the geometry of the dam: the profile of the central cantilever axis, $\eta(\zeta)$, the law of variation of the thickness with elevation, $\lambda(\zeta)$, and the law of variation of the crown radii with elevation, $R(\zeta)$. Constant-thickness arches are assumed for simplicity. All integrals are definite integrals between limits 0 and ζ. E_c, E_r are the Young moduli of concrete and foundation bedrock, respectively.

3.1 – The choice of a function $e(b,\zeta)$ gives the parametric equation of the relative excentricity $f(\zeta)$ along the central monolith under dead-weight:

$f(\zeta)=e(b,\zeta)$; constraints: $e(\zeta=0)=0$, $e(\zeta=b)=0$, $0 \le e(b,\zeta) \le 1/6$ for $0 \le \zeta \le b$, $-1/6 \le e(b,\zeta) \le 0$ for $b \le \zeta \le 1$ (b is our first parameter) (1)

Under design load the relative excentricity $f_L(\zeta)$ along the central monolith is defined by another suitable function $e_L(b_L,\zeta)$, where b_L are either equal or distinct from b : $f_L(\zeta) = -e_L(b_L,\zeta)$, with constraints:

$e_L(\zeta=0)=0$, $e_L(\zeta=b_L)=0$, $0 \le e_L(b_L,\zeta) \le 1/6$ for $0 \le \zeta \le b_L$, $-1/6 \le e_L(b_L,\zeta) \le 0$ for $b_L \le \zeta \le 1$ (b_L is our second parameter) (2)

The choice of a suitable function $T(\Lambda_0,\alpha,\zeta)$ gives the parametric equation of the section thickness $\lambda(\zeta)$ along the central monolith:

$\lambda(\zeta)=T(\Lambda_0,\alpha,\zeta)$ where $T(\zeta>0)>T(\zeta=0)$ (Λ_0 and α are our third and fourth parameters) (3)

Fig. 1 shows the relative excentricity $f(\zeta)$ as a function of parameter b.

The shape of the centerline of crown cantilever (coordinate y being positive towards upstream, and $y = 0$ for $\zeta = 0$) is denoted by:

$y = \eta(\zeta)$; then $\eta(\zeta)$ is to be determined by condition **(1)**. Defining the integral function of ζ:

$$\Lambda(\zeta) = \int \lambda(\xi).d\xi \tag{4}$$

condition **(1)** yields the following equation for $\eta(\zeta)$:

$$\eta(\zeta) = \int f(\xi).\lambda^2(\xi).d\xi/\Lambda(\xi) + f(\zeta).\lambda(\zeta) \tag{5}$$

from which $\eta(\zeta)$ is seen to be a parametric function of b , Λ , λ . The dead-weight bending moment $M_b(\zeta)$ on a generic horizontal section is then a function of η, λ, b (with γ_c = unit weight of concrete):

$$M_b(\zeta) = \gamma_c.[\eta(\zeta).\Lambda(\zeta)-W(\zeta)]_b \quad , \quad \text{where} \quad W(\zeta) = \int\lambda(\xi).\eta(\xi).d\xi \tag{6}$$

Under design load this bending moment must be offset and excentricities of opposite sign, with parameter b changed to b_L, must be generated; thus the full-load bending moment $M_L(\zeta)$ must be:

$$M_L(\zeta) = - M_b(\zeta) - M_{bL}(\zeta), \text{ a function of } b\,,\, b_L\,,\, \eta\,,\, \Lambda,\, W \tag{7}$$

From this distribution of bending moments the horizontal pressure carried by the central cantilever at full load, $p_m(\zeta)$, can be obtained:

$$p_m(\zeta) = \mathrm{d}^2M_L(\zeta)/\mathrm{d}\zeta^2 \tag{8}$$

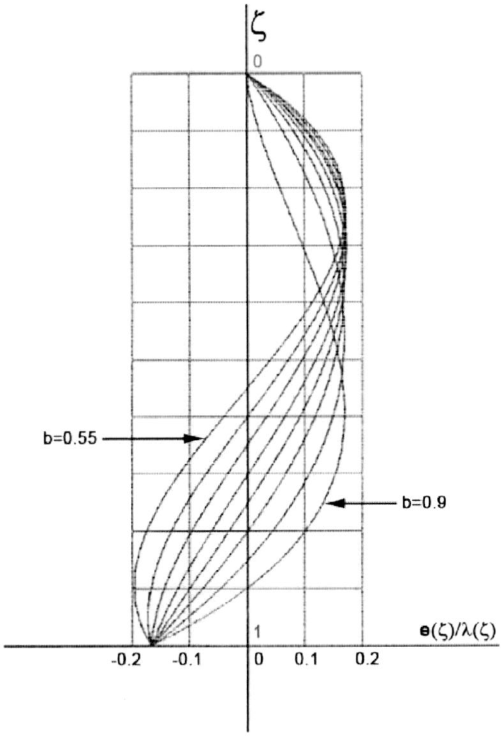

Figure 1: Excentricities.

The full-load elastica $\delta(\zeta)$ is then obtained by double integration of (7):

$$\phi(\zeta) = -12.\int M_L(\xi).\mathrm{d}\xi/[E_c.\lambda^3(\xi)] + \phi\,(\zeta=0) \tag{9}$$

$$\delta(\zeta) = -\int\, \phi(\xi).\mathrm{d}\xi + \delta(\zeta=0) \tag{10}$$

The two integration constants ϕ (ζ=0) and $\delta(\zeta$=0) are determined by the conditions of compliance of the rotation and, respectively, of the radial displacement at the interface central cantilever/foundation (ζ = 1). Now the generic arch, of crown radius $R(\zeta)$ and thickness $\lambda(\zeta)$, undergoing at crown the same radial displacements $\delta(\zeta)$ as those of the central cantilever, reacts with a pressure $p_a(\zeta)$ given approximately by:

$$p_a(\zeta) = p_{ext}(\zeta) - p_m(\zeta) \approx (8/15).E_c.\ \delta(\zeta).\ \lambda(\zeta)/\ R^2 \tag{11}$$

where p_{ext} (ζ) is the external pressure; Fig. 2 shows the standard distribution adopted (hydrostatic load plus an excess pressure distribution covering 'average' extra loads such as seismic inertial forces). From (8), (10) and (11) the last unknown, $R(\zeta)$, is obtained. Checks on horizontal (arch) stresses follow.

3.2 – Parameter fields of variation: The intervals of values within which the basic parameters are made to vary are chosen by the designer on the basis of his/her experience. As a rule of thumb the following interval limits can be suggested: $0.55 < b < 0.8$; $0.05 < \Lambda_0 < 0.15$; $0.8 < \alpha < 1.2$.

3.3 – Strategy of search: A discrete scanning of the presumed fields of variation of the basic parameters is established, giving the upper and lower bounds for each, as well as the desired number of sampling steps. Then for each possible combination of these parameters' values the computations indicated by formulas (1) to (11) under §3.1 are performed. For each parameter combination checks are moreover performed on arch stresses and the solutions exceeding the stress limits imposed as input data are discarded. Those results that are manifestly un-physical are rejected, and for each of the remaining ones the cost (or simply the concrete volume) is computed. The smaller of the latter figures, together with the quantities defining the relevant dam geometry, are retained as the 'optimal' solution. See further on Fig. 3 for details.

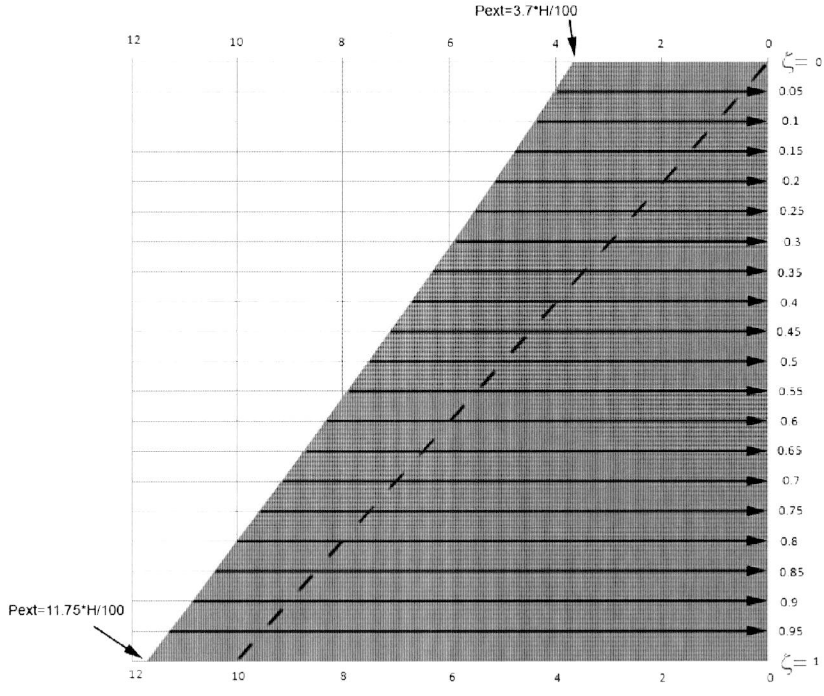

Figure 2: Standard load: external pressures (kg/cm^2) (dashed line = hydrostatic load).

4 Code structure and an example of application

Figure 3 shows the successive steps of the computational procedure, evidencing that sometimes two 'optimal' shapes with the same volume can be obtained, since the excentricity curves for dead-weight and for full load are interchangeable (except for the sign). The next Fig. 4 shows the input form, while Fig. 5 exemplifies the graphical presentation of the code output showing the results obtained for a dam 100 m high in a valley whose transverse section is depicted in the same figure. A few supplementary comments are in order:
-the 'optimal' shapes are quite sensitive to the limits imposed on concrete stresses; the designer may use this kind of information to achieve the best technical/economical balance between volume and cement content,
-the shapes obtained look adequate as a starting point for the dam final design.

5 Conclusions

5.1 - In the Authors' opinion the above outlined approach suggests that the degrees of freedom which it is advisable to adopt in the shape optimisation procedure of an arch dam can be reduced in practice – by a judicious choice of basic model and independent variables- to a very small number, probably as low as only three or four. Past research based on similar lines, though less stringently pursued, has indeed shown that clever optimization procedures based on only two to four degrees of freedom can yield shapes (and volumes) fairly comparable to those obtained by the use of much more complicated formulations, based on a high number of degrees of freedom for the shape definition and on F.E. structural modeling. In the above-abridged presentation the degrees of freedom are essentially three (or four if one considers that b can vary between the conditions of 'dead-weight on free-standing monoliths' and 'design load'):

$$a, \quad A_0, \quad b \; .$$

The basic model requires only enchained computations, avoiding the solution of huge systems of equations; thus computation times are kept small.
5.2 - From a more general point of view, the Authors think that in order to take into account the ample diversity of loads acting on the dam throughout its lifespan the definition of the optimization 'objective function' should be given more careful, critical thought. Very sketchily, in the Authors' opinion an ideal 'rational' approach could be imagined more or less as follows. The external load scenarios should be defined in detail, assigning an estimate of the probability P_L of occurrence of each scenario during the dam's lifetime. Then for each of these load sets the 'vulnerability', or the probability of failure of the dam, V_P (as an implicit function of the shape parameters) should be evaluated; lastly, the cost C_F of the consequences of each case of 'failure' should also be evaluated. Maintenance costs, C_m, and operating costs, C_e, should also be included. The total cost to be minimized, TC, could then be expressed as:

$$TC = \sum \; (C_c + C_e + C_m) + \sum_{L} \; P_L \; .V_P \, .C_F$$

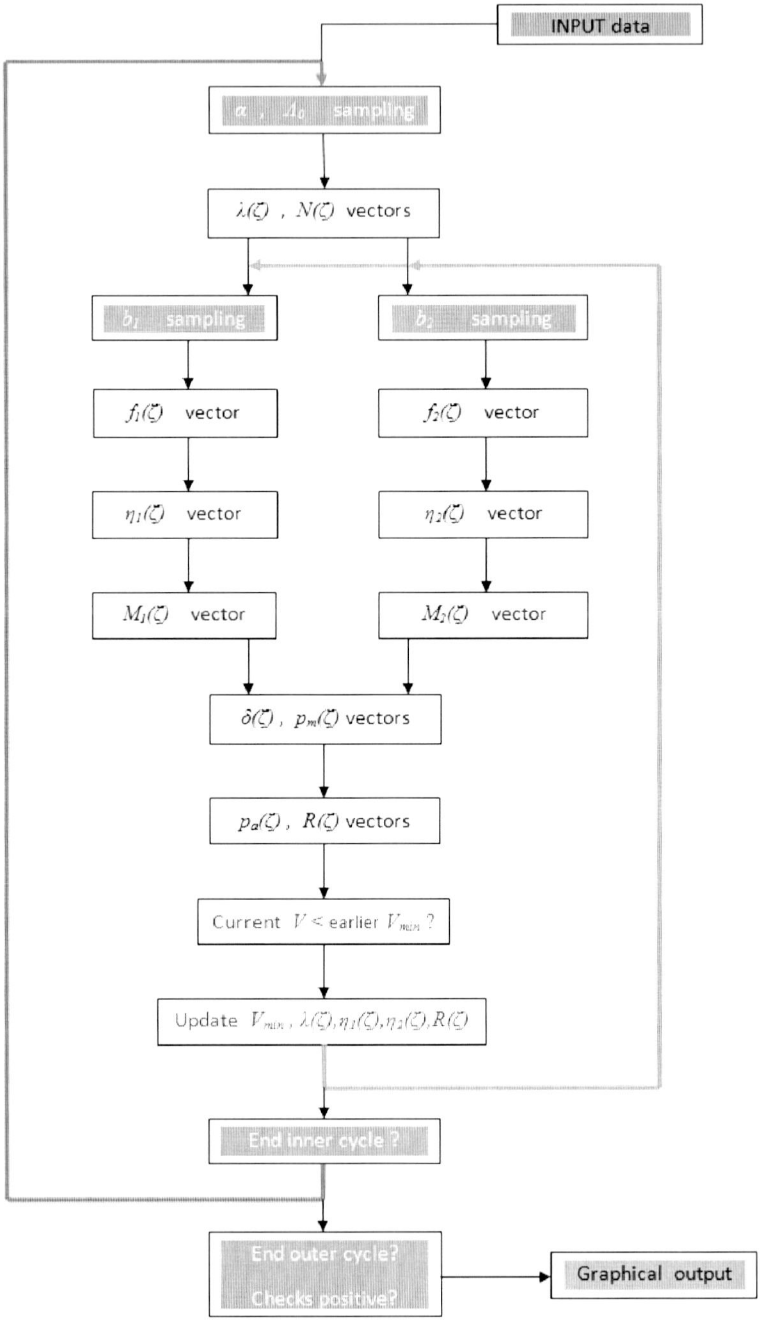

Figure 3: Flow chart of optimisation procedure.

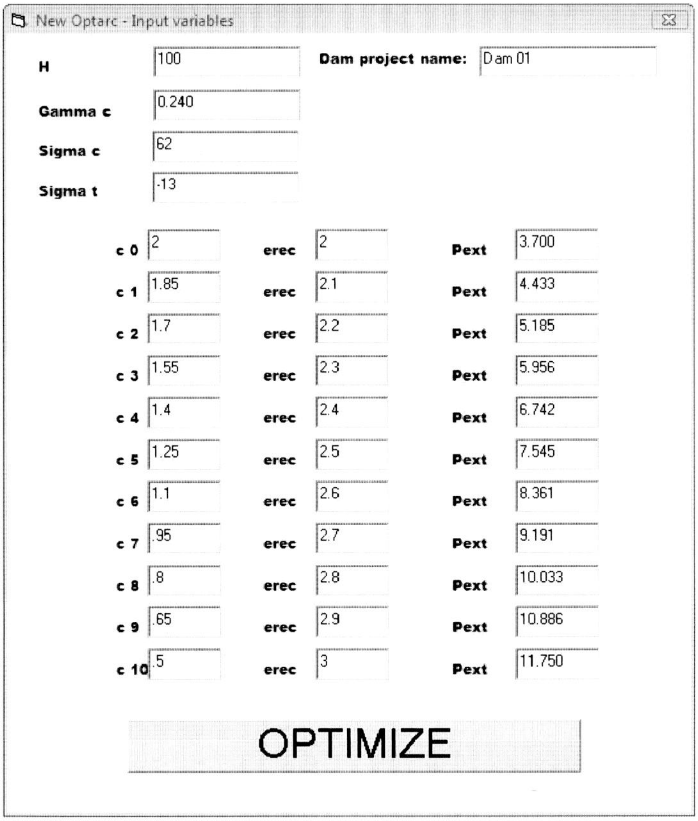

Figure 4: Input form for example of Fig. 5. The values c(I) stand for the ratio C(z)/H where C(z) is the valley chord at z meters under crest; z = I*H/10. erec = Ec/Er.

[All time-dependent costs should of course be actualised to a convenient date.] Such an objective function would be preferable to the above simpler one (§**1**), and its adoption would make reasonable the use of sophisticated models; but, alas! the outstanding difficulties of putting into practice such a 'rational' approach are all too evident.

5.3 - In the end, it goes without saying that any algorithmic 'optimisation' procedure should be followed by an 'expert' refining of the design in order to bring it into line with actual circumstances and with good practice. In this connection it is necessary that the designer's experience and intuition complement the results of the chosen algorithmic tool if an engineeringly 'safe and sound' optimal design is to be achieved.

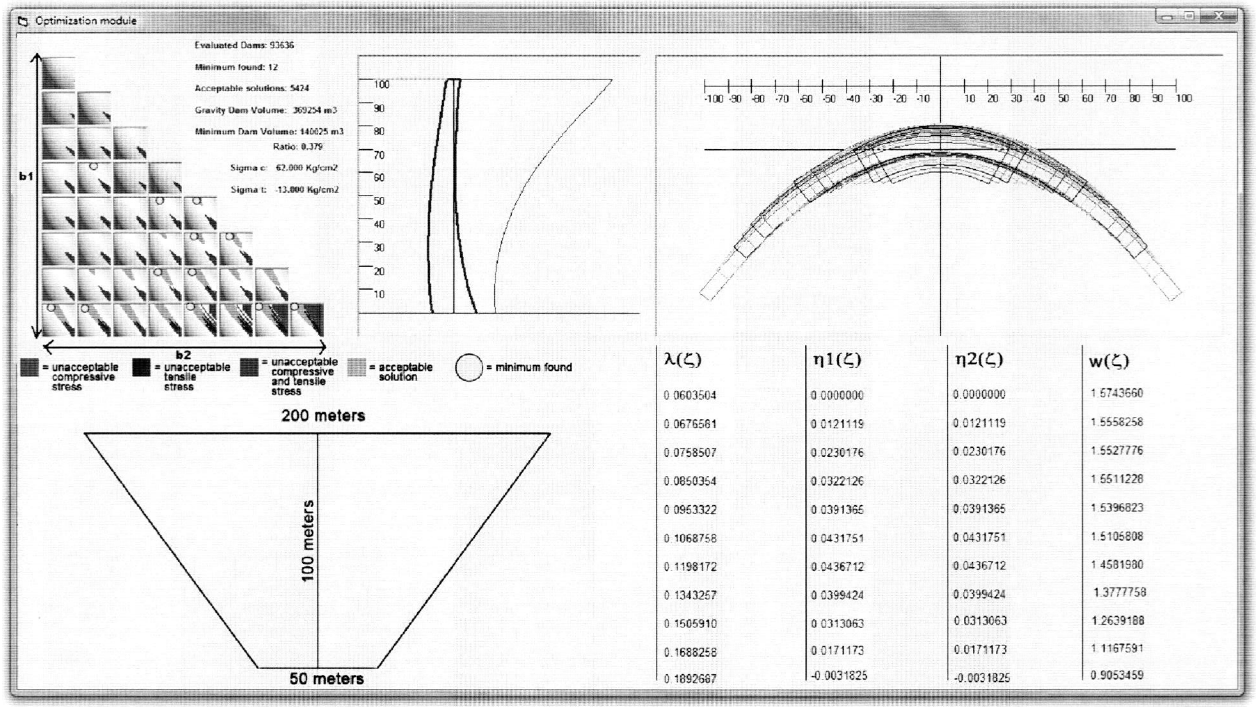

Figure 5: Examples of a graphical output for a dam 100 m high, in a trapezoidal valley with crest chord = 200 m, bottom chord = 50 m, parabolic arches, max compressive stress (Sigma c) = 62 kg/cm^2, max tensile stress (Sigma t) = -13 kg/cm^2.

References

[1] Zhu Bofang, Optimum Design of Arch Dams, *Dam Engineering*, Vol. I, Issue 2, 1990

[2] Zhu Bofang, Rao Bin, Jia Jinsheng, Li Yisheng, Shape Optimisation of Arch Dams for Static and Dynamic Loads, *Proceedings of the International Symposium on Practice and Theory of Arch Dams* – Hohai University Press, Nanjing (China), 1992

[3] M. Fanelli, Optimal Proportioning of Arch-Gravity Dams: the Automatic Generation of a Simplified Starting Geometry, *Dam Engineering*, Vol. III, Issue 2, 1992

[4] M. Fanelli, A. Fanelli, P. Salvaneschi, A Neural Network Approach to the Definition of Near Optimal Arch Dam Shape, *Dam Engineering*, Vol. I, Issue 2, 1992

[5] Orsola Brasi, Personal communication, January 2000, Metodo approssimato per l'ottimizzazione di forma delle dighe ad arco-gravità (An approximated Method for the Shape Optimisation of Arch-Gravity Dams; a thesis discussed at the Polytechnic School of Milano (referees M. & A. Fanelli)

[6] Stoyan Grigorov, Stefan Tasev and Anton Tzenkov, Extended Optimisation of Arch Dams, *Tenth Bulgarian Conference on Theoretical Mechanics*, Varna (Bulgaria), 2005.

[7] Jiazheng Pan and Jing He, Shape Optimisation of Arch Dams (Chapt. 7). *Large Dams in China*, China Water Power Press, pages 443-482, 2000.

Silo foundation optimization through reinforcement concrete shell structure

C. J. Martins[1], O. A. Figueiredo[1], L. A. Martins[2] & R. A. F. Peixoto[1]
[1]*Department of Civil Engineering, CEFET/MG, Belo Horizonte, MG, Brazil*
[2]*Geological Survey from Brazil – CPRM Regional Bureau, Belo Horizonte, MG, Brazil*

Abstract

The objective of the present work is to perform a technical and economic analysis of the adoption of shell as a structural solution for the foundation of industrial vertical silo. Initially projected in a rigid and flat slab, the foundations of a vertical silo were optimized by numerical computational modeling through a parallel Finite Element Method code, adopting shell in the form of reversed cone. It will be shown that a significant reduction of structural concrete of steel was obtained with the optimization of the silo foundations, and by this way, economy in the execution of the structure.
Keywords: non-linear analysis, constructive systems, finite element method, foundations in shell, parallel computing.

1 Introduction

Initially, the Foundation Engineering study based itself on empirical observations of the structural and ground behavior of the infrastructure elements. Over 30 years, analytic theories of soil behavior were being developed and improved, being able to be cited, among others, the works developed by Terzaghi and Frolich [1], Skempton and MacDonald [2] and Taylor [3]. With the advent and solid application of the numerical computational approaches (as by example, Finite Element Method, Finite Differences and Boundary Elements), complex models for soil-structure iteration simulation have been conducted to

WIT Transactions on The Built Environment, Vol 106, © 2009 WIT Press
www.witpress.com, ISSN 1743-3509 (on-line)
doi:10.2495/OP090211

more detailed analysis and structural designs, and by this way, for more rational foundations (Kurian [4]).

Among the several structural elements used for foundations of constructions, applicable to most soil types, the concrete floor is a method of shallow foundation that has many advantages when specified and correctly implemented. Overall, concrete floors are massive slabs that support and distribute, in a uniform way, the loads of buildings to ground. Given its simplicity of execution and relatively low tension transferred to the ground, it has been widely used in both residential (Martins et al. [5]) works as industrial (Kurian [6]).

To resist the loads from the superstructure acting perpendicular to your plan, in a flat foundation arise internal forces of bending and shear. By another way, in shell foundations (such as the hypar shells) internal forces due superstructure loads provide only internal forces in your plan, mainly tensions compression and in-plane shear (Flugge [7]). This difference in behaviour implies a structural optimization in the dimensioning of the shell foundations, since the material is subject mainly to internal forces of normal compression. The economy of material becomes more evident when such elements of foundation are made of structural concrete, whose high performance under action of internal forces of compression provides a great reduction in consumption of material.

The modern concept of using shells as structural elements of foundations is being addressed by researchers worldwide. Candela can be considered as the forerunner of art, introducing foundations of paraboloid hyperbolic shells in the soil of the city of Mexico (Candela [8]). Over the years, many works addressing studies and applications of shell foundations were developed. Kurian [9] presents a feasibility study of these foundations. Hanna and Abdel-Rahman [10] researched the behavior of conical shell settled under sand. Maharaj [11] and Martins et al. [5] analyzed the behavior of these foundations in clay, by non-linear Finite Element Method. Nicholls and Izadi [12], Melerski [13], Kurian [6] studied the bearing capacity of foundations in paraboloid hyperbolic form. Kurian and Varghese [14] and Kurian and Mohan [15] analyzed the distributions of the contact tensions of foundations in shell. Harrop-Williams and Grivas [16], Chandrashekhra and Antony [17] analyzed the interaction between foundations in non-homogeneous elastic soil. Melerski [13] developed an approached solution for the problem of the shell of small thickness resting on elastic base. Kurian [18] discusses the applicability and advantages of the use of shells as elements of foundation. The evident economic benefits from the use of shallow foundation in the form of shells are proven in several reports of applications in various parts of the world (Kurian [6]). In general this technique shows great reduction of consumption of materials and consequent increase in complexity in the work of excavation. As reported in the literature, the foundations in shell are extremely advantageous when they are submitted in at least one of the following situations: low carrying capacity of the land or cost of the workforce relatively low in comparison to the cost of materials (steel reinforcement and structural concrete).

The use of shallow foundations made in structural reinforced concrete in the form of shells, in particular the shell in chop-conical geometry, as element of foundation for a set of vertical circular silo with 30.000 t of capacity, situated in a region whose soil presents relative low bearing capacity, is proposed in this work, aiming at reducing the overall costs of the structure. For this, analyses by FEM on cluster of PCs were conducted, taking into account the non-linearity of the concrete and soil. The maximum internal forces, obtained by these analyses, were used in the design of concrete and steel reinforcement sections. Finally, the reduction in materials quantitative obtained from the adoption of the chop-conical shell is presented.

2 Foundations in conical shell geometry

2.1 Geometrical aspects

Although a variety of shells adjust themselves to be used in coverage, those ones that can be used in foundation are few. Among these, there is the chop-conical surface, whose geometric characteristic allows this shell to be used efficiently and effectively in foundations under different circumstances (Kurian [6]). The rotationally symmetrical conical foundation in reinforced concrete, shown in Figure 1(a), is probably the simplest way in which a shell can be put to use in foundations. The placement of the bars is as simple as for flat foundation, while construction is only a little more difficult. The shell may be of uniform thickness, or it may be of varying along its depth. Cones of substantially larger dimensions can also serve as foundations for tall structures such as chimneys, instead of conventional circular or octagonal rafts (Figure 1(b)). The shell of inverted position in Figure 2(a) is appropriate for structures such as guyed Masts. Large inverted conical shells, in their full form or truncated, depending on the required area of contact with the soil, can also be used like foundation for cylindrical structures such as silos and storage tanks grounded (Figure 2(b)).

2.2 Constructive aspects

The simplicity in the materialization of the surface of the foundation in conical shell is one of the main characteristics that make the use of this structure economically viable. The sequence constructive is the same that conventional flat foundation, that is: excavating and preparing the soil, positioning of steel bar, casting of concrete. In the following these steps are described. For details see Kurian [6].

2.3 Design aspects

2.3.1 Structural behavior

Internal forces developed in conical shell structures are much smaller than conventional flat foundations, since this one must resist to bending inner forces by the presence of moments. By another way, the cracks of the concrete structures are almost eliminated in the shell, reducing the risk of corrosion of reinforcement by exposure to moisture from soil. Since shell structures generally are subject only to traction and compression stress, which are always distributed at cross sections in a uniform way, resulting in better use of the materials and consequently with economy.

2.3.2 Geotechnical behaviour

In general, the geotechnical behavior of the foundations in curve structure has been analyzed in a similar way to the foundation plans structures, adopting itself to the model of Winkler-Pasternack for soil (Maharaj [11]) or through the discretization of soil (Martins et al. [19]). Experimental and theoretical studies have been conducted, whose performance of the structures in shells has presented superior performance than flat ones. Nicholls and Izadi [12] held an experimental research on shell footing with the purpose of determining the distribution of pressure of contact as a function of the ultimate load. The results indicate that ultimate bearing capacity and support displacement for shell structures presented better results when compared with their corresponding conventional flat. Kurian and Varghese [14], by experimental results, confirmed that the use of shell elements leads to more economic structures. Hanna and Abdel-Rahman [20] studied the geotechnical behaviour of shell footings placed over sand and under conditions of axial loading. The results show the admirable performance of foundations shell with respect to ultimate bearing capacity and displacements. In this work, the numerical modelling of the behaviour geotechnical for shell structures was conducted considering soil as a perfectly elastoplastic material and Mohr-Coulomb criteria, considering a parallel implementation of FEM code. As it will be seen, the geotechnical behavior of shop-conical shell presented results seemed flat one.

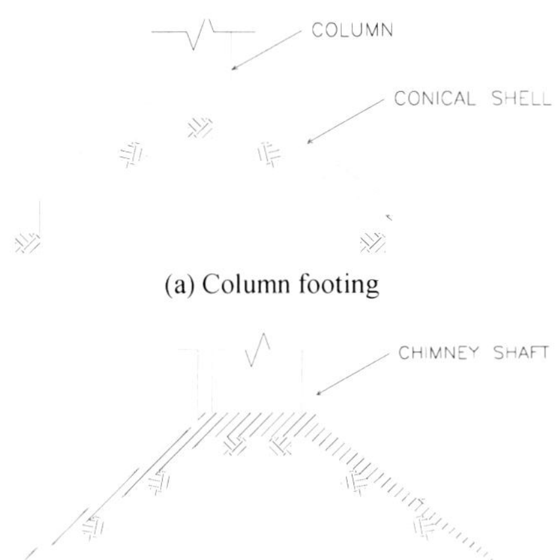

(a) Column footing

(b) Chimney raft

Figure 1: Conical shell foundations.

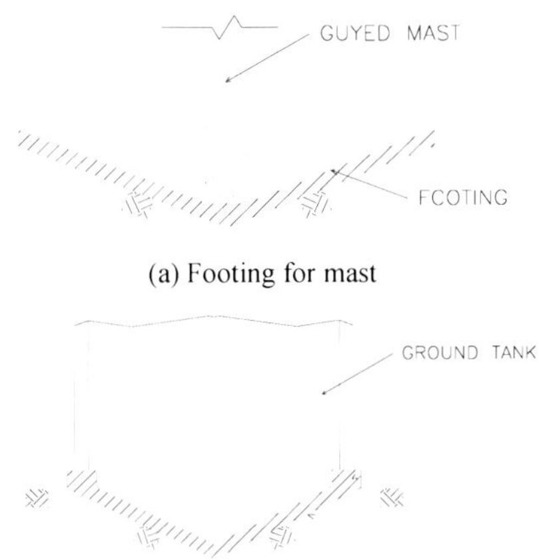

(a) Footing for mast

(b) Raft for ground tank

Figure 2: Inverted cone foundations.

WIT Transactions on The Built Environment, Vol 106, © 2009 WIT Press
www.witpress.com, ISSN 1743-3509 (on-line)

3 Numerical modeling by FEM

The finite element method (FEM) is one of the most widely considered computational tools used in solving physical problems. Initially developed aimed at analyzing problems of structural engineering Aeronautics is currently widely used in all areas of Science. The basic idea of numerical analysis by FEM is the subdivision of the field of physical problem with subdomains where the behaviours of the problem is easily represented. It is not intended to calculate the displacement of all points in the subdomains of the structure, such as analytical solutions. In a first instance are calculated only the field variable of some points which are named the nodes of model. Thus, the analyses in these individual subdomains are made and then they are interconnected, simulating the physical problem as a whole. This way it is performed an approximation of the physical response of the real problem, which is more or less close to the exact solution, depending, among other factors, the number of subdomains used in the representation of the physical problem.

One of the reasons for which the method of finite elements succeeded since the beginning of its formulation until today is that its basic concept, the subdivision of the area, produces many simultaneous algebraic equations, which are generated and resolved with the help of digital computers. Thus, you can use standard procedures, for subdomains, which do not involve engineering decisions during the computational procedure. All decisions are taken by the analyst in the stage of assembling the model, before beginning the analysis, choosing the size, arrangement and appropriate element that represents a given physical situation. Hence the idea of the element is very well defined, that is, their mathematical formulation adequately represents the physical behaviour that is intended to simulate.

The numerical modelling via FEM of concrete structures and the layers of soil under the building, were model considering material perfectly elastoplastic (Mohr-Coulomb criteria, Lemaitre and Chaboche [21]). For the FEM analyses that have been conduced here, since that discretization of soil leads to analyses of large degree of freedom, have been used up resources of parallel processing through cluster of PCs, considering paradigm of domain decomposition.

4 Industrial application

With the objective of optimizing a vertical circular silo foundation (capacity of storage 1, diameter of 1 and height 1), initially designed like flat plan concrete structure, was proposed the adoption of a conical shell foundation. The definition of the geometry of this shell was conducted from a FEM serial numerical modeling by means of commercial software, simulating soil behavior like elements of spring (Winkler's model). Figure 3 presents geometry of silo, taking into account the flat plan foundation initially designed (Figure 3(b)) and the shell foundation proposed in this work (Figure 3(a)).

WIT Transactions on The Built Environment, Vol 106, © 2009 WIT Press
www.witpress.com, ISSN 1743-3509 (on-line)

From the geometry defined initially (Figure 4) and to the appropriate forecast of geotechnical behavior for soil-silo interaction, the structural system was modeled using computational code based on 3D parallel FEM elastoplastic, using prismatic Lagrangian elements of 21 nodes, with a maximum size of 0.4 meters. The simulation of the ground as a semi-infinite medium was obtained by truncating the model is 16 meters deep and 150 meters at horizontal directions.

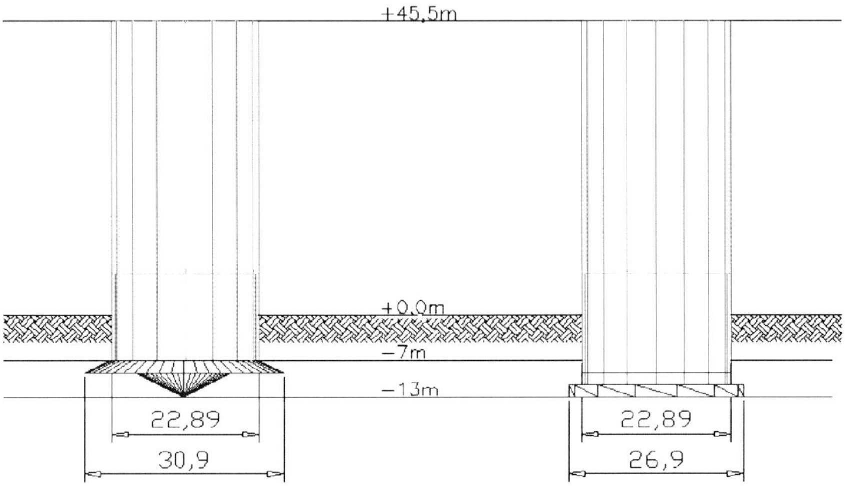

(a) foundation in shell (b) flat foundation

Figure 3: Geometry of the granary and foundations (units in meter).

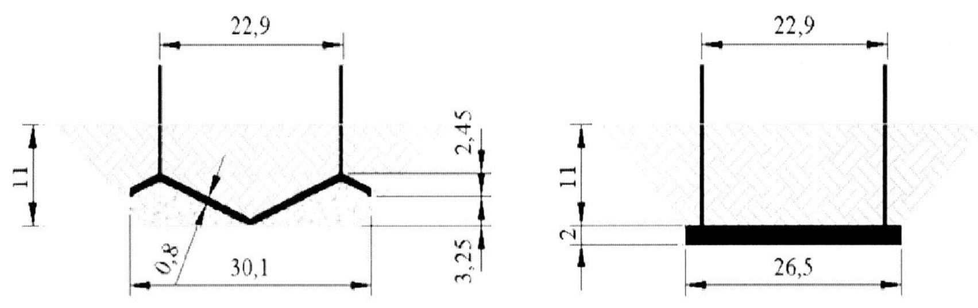

(a) foundation in chop-conical (b) flat foundation

Figure 4: Cross-section of the foundations (measures in meter).

Concrete class C30 with, with fck=30MPa, initial modulus of elasticity E=30GPa and Poisson coefficient μ=0,20 was adopted. The steel used in reinforcement of the structures in concrete, CA-50 steel, has characteristic resistance fyk =500MPa and elasticity module equal to 196GPa. For the determination of concrete and reinforcement sections was adopted standard Brazilian code NBR-6118 [22]. It was considered reinforcement cover equal to 3cm for foundation, opening of cracks limited to maximum value wk = 0.2mm. The soil, as indicated by geotechnical investigations conducted on site, consists of an initial layer of 3 meters thick composed of moderately compacted silt-clay, followed by a layer of soft silt-clay with 6 meters thick. Under this last layer, lies a layer in clay very hard and compacted. The level of the water was not found.

Figure 5 presents the mesh considered in the simulation of the semi-infinite environment. The discretization of model was made by approximately 6 million of 3D elements, totaling up 204 million degrees of freedom. The analyses were conducted considering parallel processing in PCs cluster according domains decomposition paradigm. The entire mesh was partitioned in nine subdomains, that were executed in independent CPUs, adopting the Conjugated Gradient with Jacob Precondition (PCG) algorithm in the resolution of the resultant algebraic system. In the Figure 6 and 7 are presented the results of vertical displacement of the foundations (post-processed through commercial software in the viewing of the results), obtained from the numerical analyses performed in cluster of PCs. Table 2 presents a summary of the quantitative involved in the execution of the foundations, as well as the results of foundation displacement.

5 Conclusions

The results obtained of the case study (Table 2), composed by the foundation of a cylindrical vertical silo with 30.000 t, showed that the consumption of steel was reduced by 23% of the initial value while the consumption of structural concrete was reduced to about 60%. Such difference does not cause surprise; therefore while a foundation plan element suffers efforts of bending moments when subject to a vertical load, including its self-weight, the foundation in conical shell maintains the load applied, primarily of necessities related to the behavior of shell (compression and traction or shear, all they acting inside the plan of the surface). In the conical structures, even when present, the efforts of bending assume barely values of secondary importance.

Table 1: Variation of the physical properties of soil with the depth.

Depth (m)	NSPT	Module of elasticity (MPa)	Coefficient of Poisson	Angle of internal friction	Cohesion (MPa)
1	28	7,00	0,35	45	0,20
2	28	7,00	0,35	45	0,20
3	34	8,50	0,30	50	0,30
4	4	1,00	0,40	40	0,20
5	5	1,25	0,40	40	0,20
6	8	2,00	0,40	40	0,20
7	11	2,75	0,40	40	0,20
8	10	2,50	0,40	40	0,20
9	12	3,00	0,40	45	0,20
10	16	4,00	0,30	45	0,20
11	24	6,00	0,30	50	0,03
12	27	6,75	0,30	50	0,03
13	36	9,00	0,30	50	0,03
14	40	10,00	0,30	50	0,03
15	57	14,25	0,30	50	0,03

Clearly, any structure will take shape only when the construction materials are combined with the workforce. Therefore the economy about the material reduction should be confronted to the increase in the quantity of labor and/or necessary equipment for excavation and landfill. It should be remembered also that even when the total quantity of excavation for execution of the foundations in shell be equal or even smaller than for flat structures, the unit cost of the work by quantity of material is bigger for the shell structures, in function of the its geometrical complexity. Taking into account the two factors mentioned above, it can be affirmed that the adoption of the cone-reversed shell surface will be economically viable whenever the economy about the cost of the material (concrete and steel) exceed the extra cost by the earthwork.

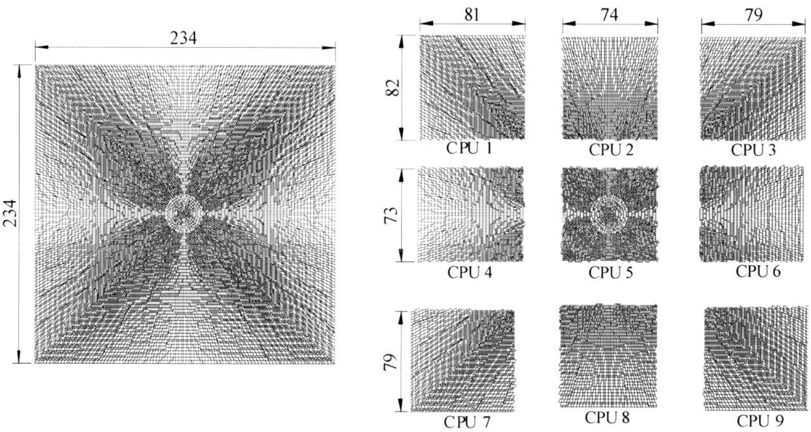

Figure 5: Upper sight of the mesh FEM and Domain Decomposition.

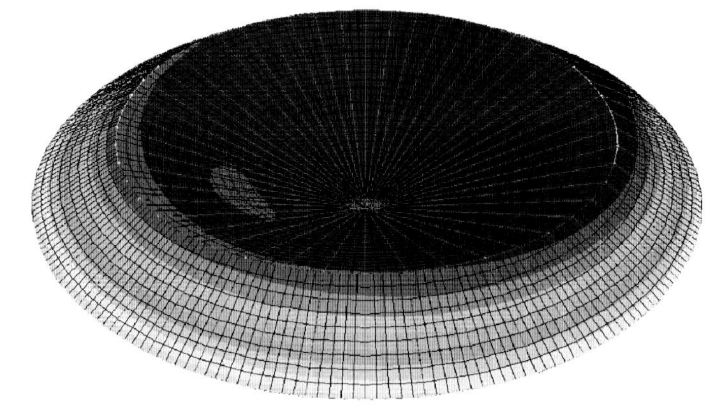

Figure 6: Foundation in shell: vertical displacement (measured in millimeters).

In relation to the differential displacement, since plan foundation is very rigid it does not present this type of deformation while conical shell suffered 9 mm, corresponding to an angular distortion of about L/30000, value with negligible impact on performance of the silo. Finally, in this work it can be shown with no doubts that the adoption of the inverted cone led to reduction in overall structure cost, because the reduction in consumption of inputs (concrete and steel) is certainly greater than the increase in consumption of labor.

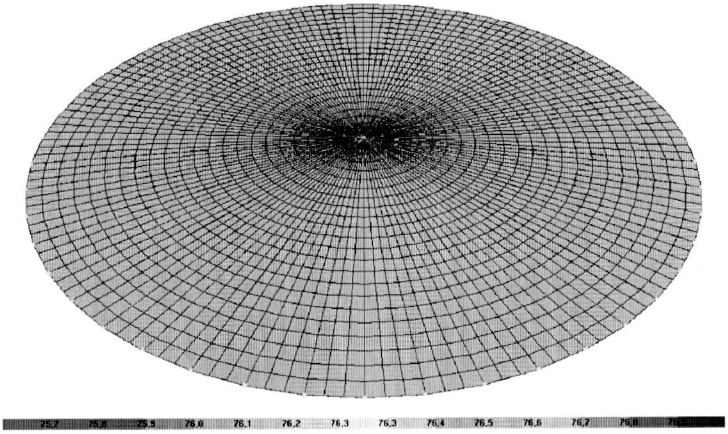

Figure 7: Flat foundation: vertical displacement (measured in millimeters).

Table 2: Comparative chart between proposals.

Foundation in shell	Flat Foundation
Concrete C30 = 671,9 m^3	Concrete C30 = 1136,6 m^3
Reinforcement = 705,7 t	Reinforcement = 2973 t
Forms = 77,8 m^2	Forms = 170 m^2
Cushion concrete = 83,9 m^3	Cushion concrete = 56,8 m^3
Excavation = 2443 m^3	Excavation = 3175 m^3
Landfill = 1771 m^3	Landfill = 2495 m^3
Maximum displacement = 71 mm Differential displacement = 9 mm	Maximum displacement = 76,4 mm Differential displacement = null

Acknowledgement

The authors thank to the FAPEMIG for the financial support.

References

[1] Terzaghi, K. and Frolich, O.K., 1936. Theorie der setzung von tonschichten, F. Deuticke.

[2] Skempton, A.W. and MacDonald, D.H., 1956. The allowable settlement of buildings, in Proc. of Institute of Civil Engineering, 5, Part III.

[3] Taylor, D.W., 1948. Soil mechanics, Willey International Edition, John Willey e Sons.

[4] Kurian, N.P., 1982. Modern foundations-Introduction of advanced techniques, Tata McGraw-Hill, New Delhi.

[5] Martins, C. J.; Figueiredo; O.A., Peixoto, R.A.F.; Martins, L.A , 2008. A Study of Foundations in Low Bearing Capacity Soil Using a Hyperbolic Paraboloid Shell. In: The Sixth International Conference on Engineering Computational Technology, Athens, p. 1-16.

[6] Kurian, N.P., 2006. Shell foundations – Geometry, analysis, design and construction, Alpha Science International Ltd.

[7] Flugge, W., 1973. Stresses in shells, New York Springer-Verlag.

[8] Candela, F., 1955. Structural applications of hyperbolic paraboloidal shells, Journal of the American Concrete Institute, vol 26, 5, 397-415.

[9] Kurian, N.P., 1977. Economy of hyperbolic paraboloidal shell footings. Geotech. Eng., 8: 53-59.

[10] Hanna, A.M. and Abdel-Rahman, M.M., 1990. Ultimate bearing capacity of triangular shell strip footings on sand, Journal of Geotechnical Engineering, ASCE, Vol.116, No.12 pp. 1851-1863.

[11] Maharaj, D.K., 2003. Nonlinear Finite Element Analysis of Strip Footing on Reinforced Clay, Electronic Journal of Geotechnical Engineering, Vol. 8, Bundle C, Paper 2003-0338

[12] Nicholls, R.L. and Izadi, M.V., 1968. Design and testing of cone and hypar footings, Journal of Soil Mechanics and Foundation Engineering, ASCE, Vol. 94, No. SM1, pp. 47–72

[13] Melerski, E., 1988. Thin shell foundation resting on stochastic soil, Journal of Structural Engineering, Vol.114, No.12, pp. 2692-2709

[14] Kurian, N. P. and P. C. Varghese, 1969. Discussion of "Design and testing of cone and hypar footings," by D. L. Nicholls, and M. V. Izadi. Journal of Soil Mechanics and Foundation Engineering, ASCE, 95(SM1): 415-416

[15] Kurian, N. P. and C. S. Mohan, 1981. Contact pressures under shell foundations, Proceedings of the 10th International Conference on Soil Mechanics and Foundation Engineering, Stockholm, Sweden, Vol. 2, pp. 15-168

[16] K.O. Harrop-Williams and D. Grivas, "Interference between geotechnical structures", Journal of Geotechnical Engineering, vol. 111, no.3, pp. 412-464, 1985

[17] Chandrashekhra, K. and Antony, S.J., 1996. Interaction analysis of strip footing resting on a non-homogeneous elastic medium, Computers and Structures, Vol.60, No.1, pp.79-86

[18] Kurian, N. P., 2003. Shell foundations-rationale and use, International e-Conference on Modern Trends in Foundation Engineering: Geotechnical Challenges and Solutions, 21-26 January, to be Organised by Geotechnical Engineering Division, Department of Civil Engineering, Indian Institute of Technology, Madras, Chennai-600036, India, pp. 1-18.

[19] Martins C.J., Araújo F.C., Mansur W.J., 2002. An efficient BE iterative-solver-based substructuring algorithm for 3D time-harmonic problems in elastodynamics, Engineering Analysis with Boundary Elements.
[20] Hanna, A.M. and M. Abdel-Rahman, 1998. Experimental investigation on shell foundations on dry sand. Can. Geotech. J., 35: 828-846.
[21] Lemaitre, J. and Chaboche, J. L., 1994. Mechanics of solid materials. Cambridge University Press.
[22] NBR-6118, 2003. Projeto e execução de obras de concreto armado, ABNT.

Optimum design for reinforced concrete pavement by parallel finite element code

C. J. Martins[1], R. R. Silva[1], L. A. Martins[2], E. V. Andrade[1] & F. R. G. Padula[1]
[1]*Department of Civil Engineering, CEFET/MG, Belo Horizonte, MG, Brazil*
[2]*Geological Survey from Brazil – CPRM Regional Bureau, Belo Horizonte, MG, Brazil*

Abstract

In this work, the structural behavior of reinforced hollow-core concrete slabs considered as industrial pavement will be displayed. The geometry of the plates was obtained by numerical modeling through computer algorithm based on the parallel Finite Element Method in, considering both the soil and concrete like elastoplastic materials. It will be shown that greater speed of execution and significant reduction in consumption of materials, and by this way, economy in the implementation of an industrial pavement through the design of a precast hollow-core concrete slab was obtained.
Keywords: PCP, constructive systems, non-linear finite element method, concrete pavement, hollow-core slabs, precasting concrete.

1 Introduction

Over the past decades, the industry focused on the construction of reinforced concrete floors is increasingly applying technological advances, resulting from extensive research on this topic. Such studies, involving the development and creation of new constructive techniques, development of concepts of structural design, allied to improvements made by standards aimed at this structural material has led to concrete pavements with high performance and durability, making these structures very competitive (Nasir et al. [1]).

Among the various technological developments for the construction of concrete railways, the adoption of precast concrete plates has made great

WIT Transactions on The Built Environment, Vol 106, © 2009 WIT Press
www.witpress.com, ISSN 1743-3509 (on-line)
doi:10.2495/OP090221

progress over time. The use of reinforced concrete precast panels have as main advantages the quality and speed in implementing of the pavement, and allow reduction in the amount of material, equipment and labor deployed in site. Since the front work will have as main function simply placement of concrete slabs at correct place, while another very important jobs such as bar placement, concrete casting and cure, which stages a long time and significantly affect the quality of the structure, can be conducted on a more adequate place (Bruggeling and Huyghe [2], Wilde et al. [3]).

The performance of reinforced concrete pavement can be increased by adopting the concept of the pretension (Lin and Burns [4]). The introduction of compression prior to live load may be obtained by a rational way if it will be allied to the concept of precast. The pretension leads to panels with larger extensions and hence lower density of joints, prevention of premature cracks in the concrete and sections more economical (Corley et al. [5]).

Although the concept of prestressed concrete been widely applied in the construction industry in superstructure elements, for structural pavement few technical developments are reported. Hargett [6] reported the construction of a constructive system composed of precast concrete deck supported by a granular sub-base layer covered by sand. On the plates of concrete is applied layer of asphalt concrete. Also according Hargett [6] that system proved to be inefficient, since constructive difficulties caused by connection between the plates and premature cracks occurred in the asphalt layer. In Nishizawa et al. [7] is described a system consisting of paving slabs in precasted concrete of variable dimensions in plan, supported by mechanically stabilized sub-base, which showed satisfactory results. Kumakura et al. [8] proposed a system for load transfer between concrete slabs, composed of prestressed steel bars every 0.5 meters. Through this mechanism it was possible to obtain full transfer of loads between panels and precise union between them. Chisato et al. [9] reported research involving the study of the influence of the elements of transfer of load on pre-manufactured panels with varying dimensions. From extensive research on the performance of different systems pre-existing manufactured by the 80s, Cable et al. [10] developed new and important concepts involving the design and construction of prestressed concrete pavements. In Merritt et al. [11] presents a new constructive system composed of prestressed plates in situ. This system was initially tested in the laboratory and then in real conditions of use, giving excellent results (Merritt et al. [12]). Fort Miller [13] presents a system composed of prestressed plates aimed to apply in situations of heavy traffic (Lepree [14]) and when speed in the pavement execution was necessary. Chang et al. [15] present an extensive comparative study between the main techniques of precast pavement systems in reinforced concrete, analyzing the technical and economic features of these systems, both in the execution of repairs and in the development of new paved roads.

The techniques of pretension and precast developed for reinforced concrete structures made possible the implementation of structural elements of high quality and competitiveness. Aiming to make the structural components lighter and economical, the industry of precast slabs have adopted so-called prestressed hollow-core slabs as a rationale solution for implementation at floors and bridges. The slabs are formed by panels of cross section with constant height and longitudinal hollows, which reduce the consumption of material and weight of the structure. They are produced usually by adopting high-strength concrete (Class C40 or greater) and low slump, vibrating at high frequency equipment. For their constructive technique, they have reduced material consumption, high quality, competitive cost, reduced time to assembly and enable the adoption of large spans.

Although widely applied to structures of floors of high buildings and bridges (Bruggeling and Huyghe [2], Lin and Burns [4]), there aren't any research about the use of hollow-core precast concrete panels reported in technical literature about road pavement. This work aims to investigate the structural behavior for a structural system composed of hollow-core plates of prestressed and precast concrete pavement (PPCP), under moving load due traffic of off-road trucks about 652 tons. Numerical modeling conduced by a parallel Finite Element Method (FEM) code will be performed, with de goal to obtain responses of tension and displacement for the plates. Will be shown that this system offers excellent structural performance due heavy moving loads and allow implementation of large expanses of pavement in a quickly way with competitive cost.

2 Structural concrete pavement

2.1 Concrete pavement

During the design of a structure must be evaluated several ways, with the goal to determine the best solution for the project studied. The main factors influencing the decision are: availability of materials, equipment and of labor, safety, durability, economy and time. In order to choose the best option, you need the broad knowledge of the various existing technologies. Currently, the use of concrete on pavement of roads has become an excellent alternative to asphalt one. In the concrete pavement, the layer of rolling also works as structure, redistributing efforts and reducing the stress imposed on the sub-base, which usually is the only layer performed between the concrete slabs and sub-grade.

One of the main advantages of the surface of Portland cement concrete is durability. The concrete does not deteriorate with the action of rain, oil or fuels and not skew the point of forming the tracks of wheels. Thus, the ground can work hard over 20 years without interventions to undergo repairs or maintenance.

A well-prepared project contributes to the advantages presented by the concrete pavement. A conservative design may prevent the implementation of pavement by increasing their cost so much. An underestimate can cause structural problems and reduce the life of the pavement, forcing the execution of repairs and reinforcements that increase costs and cause inconvenience to users. Currently design of structures should incorporate concepts such as durability and performance. These concepts also apply to decks. A project must be done not only to ensure safety against fracture, but to meet the conditions of service for a certain period of time.

2.2 Prestressed concrete pavement

The concrete prestressed floor is used especially when there is heavy traffic. Its main application is the tracks and yards of airports and highway. According to Schmid [16], at prestressed concrete floor, the tensile strength of concrete is controlled by previously pretension that compresses the concrete, creating in him a reserve of tension that allows a reduction in the thickness of the plate. The plate is almost impermeable and with no cracks. The prestressed floor still has the advantage of the joints can be spaced up to 150 meters, but the execution is more sophisticated. The first pavement built with prestressed concrete in Europe and the United States, dating from 1945. In the 60s, this solution was used in the construction of industrial floors at Australia. As an example it may be cited the Brazilian International Airport of the Galeão, which was built with the technology of post-tension.

2.3 Hollow-core concrete pavement

The main advantages of precast systems are the reduced use of moulds, the diversity of types, the capacity to reach large spans and its economy. The hollow-core precast floors are used extensively for all types of construction, not only for precast concrete structures, but also in combination with other materials, for example steel structures, concrete cast on site, and so on. The choice of floor system varies for each type of construction and from country to country, depending on the transportation, facilities for assembly, market availability, constructive culture and so on.

Choosing the most appropriate elements for the floor is defined by a number of intervening factors: availability of the market, availability of transport, ease of assembly, cost of services. Hollow-core slabs systems are very suitable for large spans and high loads. Besides, being lighter than the massive slabs, it is possible to find mounting equipment in an easy way. In the same context, also the speed of implementation can play an important role. It is noteworthy that in countries where the cost of labor is low, there is less need to use industrial slab systems.

3 Structural models

For the determination of deformation and stress in any kind of structure, it required the use of physical and mathematical models representing the behavior of materials and the structure itself. Due to the complexity of the rheology of the materials and geometry of the structure, are adopted simplifications that allow the solution of the problem. To determine the efforts that direct or indirect loads are acting over concrete slabs supported by the ground, several theories were developed. The methods of scaling of concrete pavements are based on the equations developed by Westergaard [17], which published articles on determination of displacements, stresses and bending moments in thin elastic plates, supported on dense liquid. Other authors also proposed equations for the purpose of establishing values for the efforts in concrete slabs, especially the graphs of Pickett and Ray [18].

The concrete pavements has the characteristic of absorbing the moments under the load and transmit to the subgrade efforts of low intensity, not requiring therefore that bearing capacity of foundation be very large. Although not requiring great strength of soil, the roads can present problems when built on expansive soils, or on thick layers of soft clay. The occurrence of abrupt changes in the characteristics of subgrade can also impair the performance of the pavement. There are several ways to model the soil as a support layer for plate's structures. The most common are:

3.1 Winkler's model

The dense liquid model consists of a series of independent springs with constant k. The tension at a point depends on the deformation at the point and is not affected by tensions in neighboring regions. The constant of proportionality k is known as the spring coefficient, Winkler's coefficient or subgrade modulus.

3.2 Linear elastic model

In linear elastic solid model, the soil is characterized by its elastic properties. While we do not exactly represent the behavior of the soil, the model is closer to reality compared to the Winkler's model. The model of linear elastic solid considered, in addition to normal reactions, shear efforts between the particles of the material. The displacement at a point is influenced by the tension at the point and the tension at points close together, this influence being inversely proportional to distance. The solid model of linear elastic analytical solution is more complex, issue that has lost importance due to advances of computer programs.

3.3 Non-linear elastoplastic model

Currently, available algorithms are developed through processes such as numerical finite elements, finite differences and boundary elements, which facilitates the application of the theory of nonlinear plasticity (Lemaitre and Chaboche [19]), for determination of structural response under several geometries and loading acting at different positions on the plates. Can be adopted various models for the soil and thickness of the plates can be variable.

4 Industrial application

A system for ore discharging consists of a floor over which off-road trucks operate. This floor has fifteen slabs of concrete structural placed over an existing concrete wall with sixteen meters high (Figure 1). These plates have dimensions of 12x4 meters in plan and are partially supported in the existing wall, resulting in a cantilever with 2.75 meters (Figure 2(a)).

The ore discharging system consist a dense and operating industrial area were the execution of concrete works are not feasible, so the slabs that form the operating floor must be performed outside the final position and subsequently transported. In addition, on these floors is acting off-road trucks of 652 tons weight, which requires high capacity of panels.

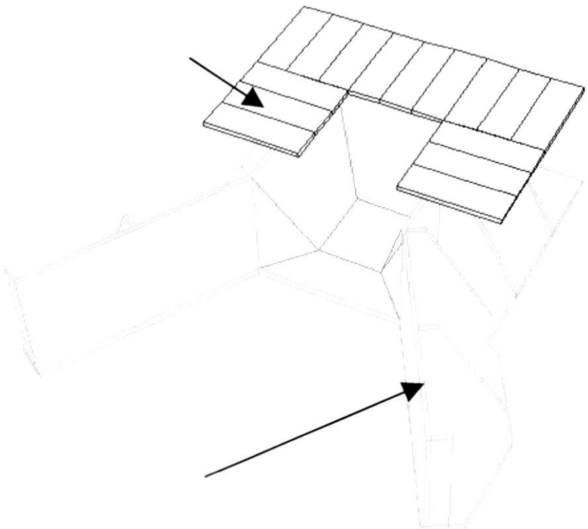

Figure 1: Ore discharging geometry.

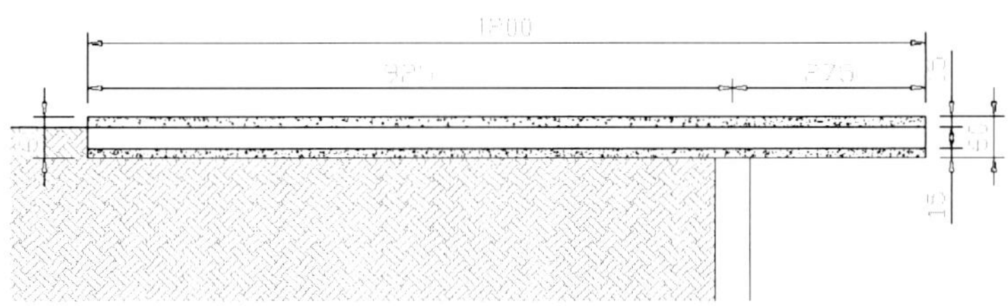

(a) Longitudinal section

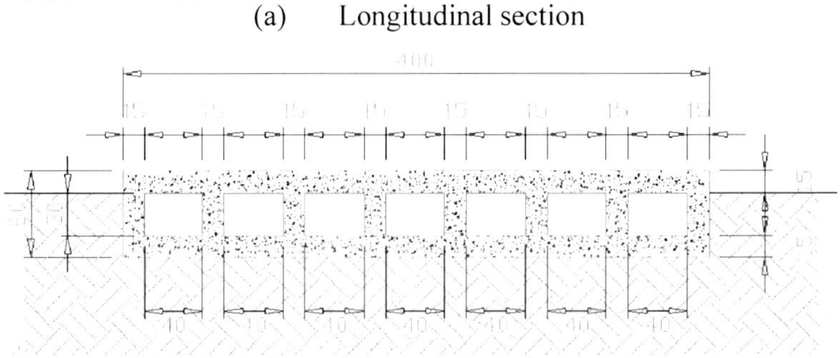

(b) Transverse section 1

Figure 2.

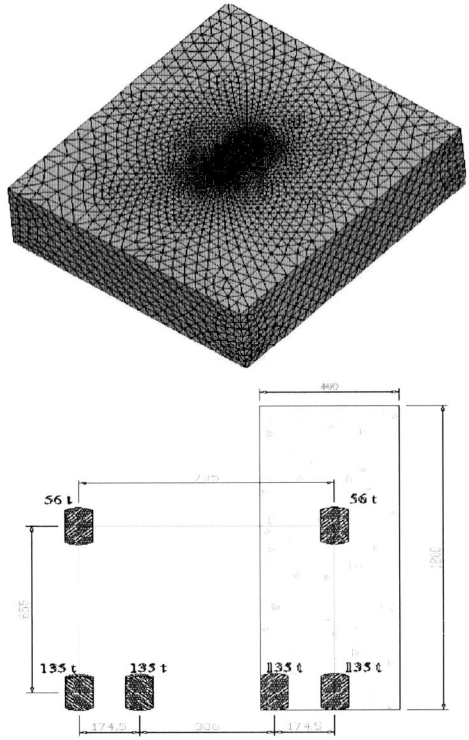

Figure 3: (a) Superior view of 3D mesh; (b) Traffic load model (centimeter, tons).

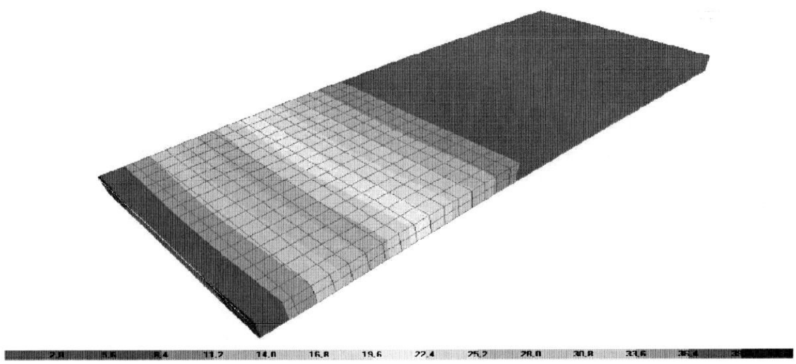

Figure 4: Vertical displacement (milímetro).

It was used to assemble precast panels hollow-core prestressed concrete since these elements combine high structural performance and lightness, allowing its transport and quick assembly. Figure 2(b) shows transversal section of the plates, made by structural concrete class c40, with fck= 40MPa, initial modulus of elasticity Ec=35GPa and Poisson coefficient μ =0,20. The steel CA-50 used for the passive reinforcing of concrete structures has resistance characteristic fyk=500MPa and modulus of elasticity Es=196GPa. For pretension was adopted CP-190RB12,7 steel, with tensile strength fy=187,3kN and modulus of elasticity Ea=195Gpa. The soil was formed of an initial layer of 16 m thick composed of compacted embankment (Ee1=1MPa, Poisson=0,4, angle of friction=15°, cohesion= 0,4 MPa) followed by a layer of compact sand thickness 6 meters (Ee2=6MPa, Poisson=0,45, angle of friction=30°). Under the compact sand, there are a layer in very hard sand. The level of the water table was not found. For the determination of reinforcement bar and concrete sections of the panels, Brazilian standard NBR-6118 [20] and NBR-7197 [21] was adopted, with concrete cover d=3cm and cracks opening wk=0,2mm.

Figure 3(a) shows the mesh of the model considered in the soil simulation like a semi-infinite space. The model was discretized by approximately 0,6 million 3D Lagrangean elements, totaling 26 million degrees of freedom. The analyses were conducted by a parallel processing according domain decomposition concept, where the mesh was partitioning in four subdomain, which were processed in independent CPUs, adopting the Conjugated Gradient with Jacob Precondition (PCG) algorithm in the resolution of the resultant algebraic system. In Figure 3(b) it is shown the traffic load model considered. Figure 4 is presented the displacements envelopment obtained from the numerical analysis, considering the traffic load model operating in several critical positions on the panel. Figure 5 presents the resulting envelopment for the tensile stress at panel. From envelopment of tensions were determined both the sections of passive bar reinforcement (0.27 cm^2) and pretension bar (7 bars CP-190RB12,7).

5 Conclusion

Techniques of prestress ally the concept of precast of structural elements in reinforced concrete was applied to industrial pavement with the purpose of reducing costs and increasing construction speed. As the constructive system was formed by hollow-core slab panels, although widely applied to structures of bridges and buildings, no application concerning paving has been reported in technical literature.

This paper presented a case study, in which precast panels formed by hollow-core prestressed panels were adopted as the rational solution for construction of industrial pavement where trucks with very high weight will operate. This structural solution in the industrial area with high traffic volume, presented themselves as the logical choice because it allowed the manufacture of boards outside the region of high-density operations.

Numerical modeling was performed by algorithm based on parallel Finite Element Method (FEM), resulting in responses of tension and displacement of the panels. From these responses, the design was conducted according to applicable Brazilian standards. It was shown that these plates allowed the construction of a pavement system with the purpose of receiving high loads of trucks, and thus absorbs very high efforts. Finally it was shown that this system offers excellent structural performance by moving loads and allows implementation of large expanses of pavement quickly and rational, may be better studied and possibly applied to highways.

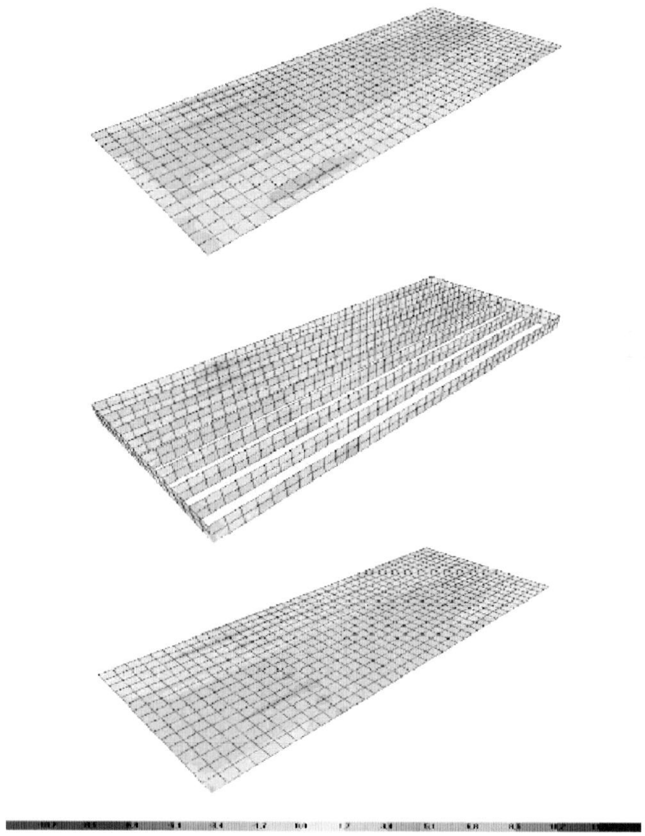

Figure 5: Tensile stress envelopment –(kPa).

Acknowledgement

The authors acknowledge the financial support by FAPEMIG.

References

[1] Nasir G.G., Joseph S., Michael ID. Evaluation of Concrete Pavement Construction Scenarios Under Performance-Related Specifications, *Transportation Research Record: Journal of the Transportation Research Board*, vol. 1813, 2007.

[2] Bruggeling A.S.G. and Huyghe G.F. Prefabrication with Concrete, *A.A. Balkema Publishers, Rotterdam, Netherlands*, 1991, 380 pp.

[3] Wilde W.J., Waalkes S. and Harrison R. Life Cycle Cost Analysis of Portland Cement Concrete Pavements, *Research Report 1739-1, Center for Transportation Research, The University of Texas at Austin*, September 1999, pp. 65-95.

[4] Lin T.Y. and Burns N.H. Design of Prestressed Concrete Structures, 3rd ed., *John Wiley and Sons, Inc.*, 1981, pp. 1-30.

[5] Corley W.G., Colley B.E., Hanna A.N., Nussbaum P.J. and Russell H.G. Prestressed Concrete in Transportation Systems, *PCI JOURNAL, V. 25,* No. 3, May-June, 1980, pp. 14-31.

[6] Hargett E.R. Field Study of Performance and Cost of a Composite Pavement Consisting of Prestressed Concrete Panels Interconnected and Covered with Asphaltic Concrete, *Highway Research Record Number 239, Highway Research Board*, 1968, pp. 137–149.

[7] Nishizawa T., Noda E. and Fukuda T. Study on the Mechanical Behavior of Precast Concrete Pavements, *International Symposium on Concrete Roads, Vienna*, 1994, pp. 93–98.

[8] Kumakura M., Kondo S., Kai K., Abe Y. and Sato R. Development of a Prestressing Method for Joints of Precast Prestressed Concrete Pavement Slabs, *International Symposium on Concrete Roads, Vienna,* 1994, pp. 83–88.

[9] Chisato H., Ikeda T., Matsuno S. and Nishizawa T. Long Term Performance of Prestressed Concrete Pavements in Japan, *Purdue Conference on Concrete, Indiana,* 1997.

[10] Cable N.D., McCullough B.F. and Burns N.H. New Concepts in Prestressed Concrete Pavement, *Research Report 401-2, Center for Transportation Research, The University of Texas at Austin*, 1985.

[11] Merritt D.K., McCullough B.F. and Burns N.H. The Feasibility of Using Precast Concrete Panels to Expedite Highway Pavement Construction, *Research Report No. 1517-1, Center for Transportation Research, The University of Texas at Austin*, 2000.

[12] Merritt D.K., McCullough B.F. and Burns N.H. Precast prestressed concrete pavement pilot project near Georgetown, Texas, *Paper No. 03-2726, TRB 2003 Annual Meeting*, pp.5, 6.

[13] Fort Miller Co. Super-slab® really is a fast (and permanent) method of repair, Super-Slab Bulletin, vol. 1, issue 1, 2007.

[14] Lepree J. Taking a Tol, *Precast Solutions,* 2002, pp10~12.

[15] Chang, L., Chen, Y. and Lee, S. Using Precast Concrete Panels for Pavement Construction in Indiana, *Indiana Department of Transportation State Office Building*, 2004.

[16] Schmid, M.T. O pavimento rígido em concreto protendido. In: *REIBRAC, 38., Ribeirão Preto, 1996. Anais, Ribeirão Preto, IBRACON*. v. 1, p. 79-91.

[17] Westergaard H.M. Stresses in concrete runways of airports. *In: annual meeting of hightway research board, Washington Proceedings*, Washington, Hightway Research Board. p. 197-205, 1939.

[18] Pickett G. and Ray G.K. Influence charts for concrete pavements. *American Society of Civil Engineers*, p. 49-73, 1951.
[19] Lemaitre, J. and Chaboche, J. L. Mechanics of solid materials. *Cambridge University Press*, 1994.
[20] NBR-6118, 2003. Projeto e execução de obras de concreto armado, *ABNT*.
[21] NBR-7197, 1989. Cálculo execução obras de concreto protendido, *ABNT*.

Author Index